D0015857

EINSTEIN

Einstein

His Space and Times

STEVEN GIMBEL

Yale
UNIVERSITY
PRESS
New Haven and London

Yale University Press books may be purchased in quantity for educational, business, or promotional use. For information, please e-mail sales.press@yale.edu (U.S. office) or sales@yaleup.co.uk (U.K. office).

Set in Janson type by Integrated Publishing Solutions.
Printed in the United States of America.

Frontispiece: Albert Einstein and his wife leaving California, 1931.

Library of Congress Control Number 2014958690
ISBN 978-0-300-19671-9 (cloth : alk. paper)

A catalogue record for this book is available from the British Library.

This paper meets the requirements of ANSI/NISO Z39.48-1992 (Permanence of Paper).

10 9 8 7 6 5 4 3 2 1

For Erin and Benjamin

CONTENTS

ACKNOWLEDGMENTS

Special thanks to Larry Marschall and Don Goldsmith for suggestions to the manuscript, Deirdre Mullane of Mullane Literary for support throughout the process, my editor Jeffrey Schier and all the wonderful folks at Yale University Press, and especially Erin and Ben Gruodis-Gimbel for help with the index.

EINSTEIN

Introduction

Albert Einstein made a mistake. It was a miscalculation, plain and simple—not mathematical but political. A group of antisemitic German nationalists had attacked his theory of relativity at a public forum. The nonsense objections were rhetorically effective only because the audience, already biased against him, did not understand the theory or the grounds on which relativity could be legitimately questioned. Einstein was not afraid of controversy and was not about to back down from a fight, but if it was a rumble they wanted, it would be done right. They would follow the rules.

Many physicists, including some of Einstein's own heroes, objected vociferously to his theory. This was expected. After all, Einstein was offering an alternative to the laws of mechanics posited by Isaac Newton and still standing after three hundred years. Not only were Newton's successes legion, they were legend. One does not, and should not, reject the bedrock of a discipline easily. Einstein knew the process and was eager to engage in it: journal articles, correspondences with fellow physicists, and panels at conferences in which the technical details of the theory are considered, evidence in favor and against the theory is weighed and discussed, and extensions and objections are made on scientific grounds. He was sure he could win on a level playing field when opposing legitimate physicists, so these politically motivated clowns would be nothing but a nuisance.

In an open letter that Einstein published in the widely read *Berliner Tageblatt*, he pointed to an upcoming occasion at which a respectable conversation about relativity would occur, the annual meeting of the German Association of Scientists and Doctors in the town of Bad Nauheim. The venerable Max Planck, the most powerful and well-respected physicist of his generation, would be presiding over a discussion about relativity theory involving Einstein himself and Nobel laureate Philipp Lenard, who opposed it. Einstein figured that these cranks felt bold enough to peddle their absurdities in the comfortable confines of public meetings surrounded by allies, but the sunlight of actual scientific discourse would send them scurrying. And so he ended with a challenge: "I notice that at the scientists' gathering at Nauheim there has, at my suggestion, been arranged a discussion on the theory of relativity. Anyone who wants to protest can do so there and put up his ideas to a proper gathering of scientists."[1]

It was a trap. Using a put-up-or-shut-up taunt as the bait, Einstein was sure that one of two things would happen. Most likely, they would not show up. There was no rational way these charlatans would dare try to hold their own against actual physicists. Their absence could be understood only as a concession, a tacit admission that they were full of hot air. But the alternative was more delicious; they would actually play the game of real science according to the accepted rules. In this scenario, they would be dispatched quickly and decisively, their snake oil shown for what it is. Either way, Einstein would be rid of them. It seemed a foolproof plan.

But Einstein was a theoretician, not an experimentalist, and theories—even good ones—don't always work when applied to the chaotic messiness of the real world. There was a third possibility he had not considered: these enemies of relativity showed up en masse at the session, a loud, unruly, and disrespectful mob willing to sabotage the proceedings with their jeers and chants. They did not play by the rules, but rather they tried to co-opt the meeting and force everyone else to play *their* game. Einstein could not easily shrug them off. They were not mere gadflies, but army ants intent on destroying anything in their way.

On the one hand, this came as no surprise to Einstein, who had left Germany at fifteen, renouncing his citizenship and promising never to return to the land of his birth exactly because he detested this sort of militant, nationalistic close-mindedness. On the other hand, the brazenness was hard to fathom. It was a sign of the times and foreshadowed dangerous things to come.

Einstein had felt the sting of antisemitism as a child, but it was after World War I that he was forced to see the depth and destructiveness that came with such hatred. His colleague, the great physicist Max von Laue, was certain that the toxic mix of nationalism, militarism, and anti-Jewish sentiment was a temporary phase in Germany and that the culture would quickly come to its senses before anything truly bad happened. Other friends went in the opposite direction; for example, Nobel laureate Fritz Haber converted and dedicated his life to nationalistic German causes—especially those of the military, becoming the father of chemical warfare. But just as Einstein predicted, such conversions would never lead to acceptance of Jewish-born Germans as authentic Germans, no matter how hard they tried or how much they personally believed they were one of them. Einstein had alienated himself from the larger Jewish community, but the times forced him to realize that his heritage was an inalienable part of who he was and who he was perceived to be. He could accept that on the grounds of those who hated him, or he could claim it for himself.

He was a pessimist, and a sarcastic one at that, but never a fatalist. Things may be bad, but you had to try to make a difference. In service of his causes, Einstein had one asset from which to derive power—his fame. At one point, he said that he had a version of the Midas touch—everything he said turned to newsprint.[2] He disliked being a celebrity and the disruption it caused, but when it was useful he would use it to give voice to the oppressed—both Jewish and non-Jewish—and to stand up against those he saw as the oppressors. This made Einstein a political figure and thereby fair game in the ugly world politics creates. He saw German society tending down a dangerous path, and he spoke up publicly against the nationalism, bigotry, and militarism that would become National Socialism. But this led to condemnation, public attacks, even death threats.

Einstein's theory of relativity contended that our universe is warped, twisted by mass and energy. Albert Einstein was a massive figure who saw that not only space, but his times, with twisted values, were similarly warped. On a trip abroad to raise money for his family and for Zionist causes, Einstein's assets—his homes, his possessions, even his sailboat—were seized by his Nazi enemies. He would be forced out of his home country, which he had left as a young man decades earlier. But what he saw there worried him.

The Nazis had rejected his theory of relativity, which famously asserted with its iconic equation $E = mc^2$ that mass is a form of energy. Yet they had also begun a program to use this knowledge to create a weapon. The speed of light, c, is a very large number and when squared becomes an incredibly great number. As a result, even if just a small amount of mass is converted into energy, the result would be powerful beyond all previous conception. Harnessing such power for positive use seemed a good idea to Einstein, but it was a plowshare that could be beaten into an unspeakably horrific sword as well. If the Nazis could figure out how to create an atomic bomb, the geopolitical consequences would be beyond dire. He had created the theoretical backdrop, and now Einstein would have to be part of preventing the era of atomic warfare. Fellow scientists convinced him that the only way to stop the German program was to beat them to the punch, and so Einstein signed a letter, written by fellow physicist Leo Szilárd, which urged the creation of what would become the Manhattan Project, thereby launching the nuclear era.

At the time, he felt he had no other choice. But after hearing of the dropping of the bomb on Hiroshima, he is reputed to have said that he could burn his fingers for having signed that letter.[3] Like the physical world, the political realm curved in strange and unexpected ways. It was relativity that led Einstein out of the patent office to become a college professor. It was relativity that made Einstein famous. It was relativity that was condemned by the Nazis for being "Jewish science." It was relativity that allowed him the political voice to oppose them. It was relativity that led to the possibility of atomic weapons and Nazi dominance. It was relativity that gave Einstein the standing to do something to stop it. It was relativity that made Einstein Einstein. Albert Einstein was a cause and an effect of history, a product of both his space and his times.

1

Everything Was in Order

Albert Einstein grew up in an adolescent Germany. Otto von Bismarck united the country into a coherent nation-state only eight years before Einstein's birth. As a result of major military victories, the formerly loose-knit regions joined together into a coherent political structure with the Prussian king, Wilhelm I, as Kaiser.

The war furor and postvictory euphoria were harnessed and turned into nationalistic sentiment. The stereotypical Prussian personality—disciplined, strict, and prizing strength and bravery over intellect and cleverness—was transformed into the cultural archetype for the young country. To reinforce this, Bismarck (the power behind the Kaiser) made the *Kulturkampf*, or cultural struggle, to reclaim true German identity official policy. This movement to cleanse German society of its unhealthy elements had a religious component. Its focus was not the antisemitism that would dominate later political discourse;

rather, it was Catholicism that was portrayed as an enemy of German progress and growth in the modern world. The political power and cultural influence of the Catholic Church, Bismarck contended, had to be strictly contained. Catholics were a minority who largely populated the working and underclass. They were not the engine that would drive Germany to its modernized, industrialized future.

The area of Einstein's birth, Swabia, lay south of Prussia and was predominantly Catholic. To northerners, Swabians were country bumpkins, cheap to a fault, and not terribly bright—traits apparent in their peculiar dialect, which among other things often added the diminutive "le" to nouns, making it seem childlike to the more sophisticated and jaded Prussians.

This general sense of alienation from Prussia made life a little easier on Swabian Jews, who, similar to Jews elsewhere, still located themselves in dense pockets, especially in smaller towns. The situation was calm enough that Swabian Jews progressively assimilated into the local culture, seeing themselves as more a part of the mainstream with each generation.

Germany was ascendant. Science and technology flourished, driving rapid industrialization that created great wealth and major changes in population distribution and lifestyle. Throughout the nineteenth century there was a tremendous flight toward the urban centers and away from the traditional rural ways of life. Einstein's family was part of this movement. His paternal grandparents, Abraham and Helene Einstein, and his father, Hermann (then age nineteen), moved from the bucolic town of Buchau to the more industrial Ulm in 1866. Hermann Einstein was a gentle soul, soft-spoken and kind and with a light and playful spirit. Optimistic and mild, he was well liked but not well suited for the cutthroat world of business. Yet that is where he ended up, joining his cousin's featherbed enterprise.

In 1876 he married Pauline Koch. Strong-willed, strong-minded, and sharp-tongued, Pauline provided an almost stereotypical counterpoint to her docile husband. Although eleven years younger than Hermann, she was the dominant personality in the marriage. According to their younger child, Maja, "There was such complete harmony in character between Hermann and his wife that the marriage would not only remain untroubled throughout their lives, but would also prove to be, at each turn of fate, the one thing that was firm and reliable."[1] Coming from a family of means, Pauline was well read and had a deep love of music, thereby bringing the sort of refinement and culture that were expected to accompany a middle-class lifestyle. Hermann would regularly read aloud in the evenings from the writings of the German poet and philosopher Friedrich Schiller and Heinrich Heine, the well-known intellectual, cultural critic, and converted Jew.

According to the birth certificate, at 11:30 a.m. on March 14, 1879, Hermann and Pauline Einstein of apartment 135 B on Banhofstrasse had a child, a son they named Albert. Although the Jewish custom had long been to name a first child, especially a son, after the child's grandfather, the grandfather's name "Abraham" was not selected, but in his honor the initial "A" was kept.

This is indicative of the sort of household Albert was born into: the family did not keep the Sabbath, did not observe the dietary code, and did not belong to a synagogue, yet they did not completely surrender their Jewish identity. They were part of the Jewish community, but like many modernist Jews of the time they were not only nonobservant, but antiobservant. Hermann scoffed at the ancient beliefs and rituals as leftover superstitions that had no place in a contemporary context. In his biography of Albert Einstein, Anton Reiser notes that "[Einstein's] father prided himself on the fact that he was a free-thinker. His belief harmonized with the thought of his time, which was controlled by the philosophy of materialism.

Albert's father was proud that Jewish rites were not practiced in his house."[2]

Indeed, they were so secular that Hermann and Pauline felt comfortable sending young Albert to a Catholic school. The family moved to Munich when Albert was a year old, and he was sent to Petersschule, or St. Peter's Academy, where he was the only Jewish boy in his class.

The featherbed business having failed a couple of years before Albert's birth, the unemployed Hermann was convinced by his younger brother Jakob to join him in an electronics venture. While Hermann had some natural mathematical abilities, he had only a high school education. Jakob, on the other hand, was a trained engineer with a college degree who saw that widespread electrification was creating a rapidly expanding market to be exploited. Jakob would run the technical side of the operation, and it seemed quite natural to bring in Hermann with his commercial experience to oversee the business end. But it was not just his knowledge that Jakob needed to get the business up and running. Pauline's family was wealthy, her father having been a grain dealer with connections to the well-off gentry in Württemberg. Hermann was able to secure funding from his family, the largest portion from his in-laws, and so the family relocated in 1877 so they could live alongside Jakob.

The Einstein brothers, while both upbeat with sunny outlooks, were different in temperament. Hermann, no matter where he was, was generally content with the world around him, while Jakob was a would-be visionary, always with a new big idea, a new plan, a new foolproof concept that he was excited about. Hermann was so good-hearted that he would get swept along with his brother's enthusiasm, often without fully deliberating on the matter—behavior that was generally quite contrary to Hermann's nature, which was usually contemplative to the point of being indecisive.

Having Uncle Jakob as part of the household was wonderful for young Albert. It was with his uncle that Einstein first began to think about mathematics, as Jakob exposed him to basic notions in algebra and number theory. Mathematics was a game—solving algebraic problems was "a merry science . . . when the animal that we are hunting cannot be caught, we call it x temporarily and continue to hunt it until it is bagged."[3] His uncle exposed him to the Pythagorean theorem, which Einstein set off to justify to himself, deriving a novel proof of the classical result.

It is not surprising that Albert's intellectual passion would arise outside of the classroom. Einstein hated the pedagogical approach of the German schools. His sister recalled that "he had a rather strict teacher whose methods included teaching children arithmetic, and especially the multiplication tables, with the help of whacks on the hands, so-called 'Tatzen' (knuckle raps); a style of teaching that was not unusual at the time, and that prepared the children early for their future role as citizens."[4] The dislike of formal schooling would last his entire life, as he believed that such methods crushed the human mind and spirit instead of elevating and liberating them as education ought to do.

But this stance did not sour Einstein on learning. From a very young age he loved teaching himself, exploring and discovering intellectual areas on his own. This led to a lifelong conviction that this was how he learned best. It was, however, a belief that would repeatedly get him in trouble.

One subject that Einstein did enjoy studying in school was religion. There was compulsory religious education in Germany at the time, and Einstein's Catholic schooling meant required classes in the catechism, where Einstein was exposed to the stories of the New Testament and the basics of Catholic theology. He was taken with them and acquired a lifelong admiration for the Jesus of the Gospels and the social justice elements of Christian belief. The teachings of Jesus, with their emphasis on love and care for those who are vulnerable, stood in stark contrast to the larger cultural messages he received from not only the school but the regular military parades that he loathed even as a child.

While his fully secular parents did not mind sending young Albert to a school of another faith, they still felt that he should know he was different from the other boys in terms of his religious heritage, and so they brought in a religious relative to give Albert an education in his own traditions. They were lessons he took to heart, especially given his experiences at school. Children can be cruel to quiet kids, especially those who do well academically, and if there is a difference to be pointed out in one of their peers, it usually serves as the basis for taunting. Thus, Einstein was bullied for being Jewish, suffering both verbal and some physical abuse. This was reinforced in one lesson Einstein would later recall, when the priest teaching the religion course brought in large metal nails and explained that it was with such implements that Jesus was crucified on the cross. Einstein recalled this story differently for different audiences—in some tellings of this story, Einstein stated that the priest's intention was to instill antisemitic sentiment in the students, whereas in other cases he contended that an antisemitic lesson was not the priest's purpose, but that upon completion of the story, all eyes in the class turned to focus uncomfortably on him.

Either way, Einstein's Jewish heritage made him an outsider, and when possible he intrinsically would embrace the role as the *Einspänner*—the lone horse—especially when it meant distinguishing himself from those with power. So, when Einstein was considered to be "the Jewish kid" in a particular setting, he would then become "the Jewish kid." He would simultaneously rebel against the authorities and bullies at school and his parents by becoming, at age eight, a deeply committed, practicing Jew—keeping kosher, observing Shabbat, even making up his own psalms to God which he would sing to himself on the way to school.

The idea of worship through song was quite natural for Einstein, who attributed special powers to music. He began violin lessons when he was six, and the listening and playing of music would be a constant through all of his life. It appealed to him because it was a direct experience unmediated by language.

Many of us believe that we think in words, and that beliefs are sentences that we take to be true of reality. Language is the means we use to make sense of the world. Every object has a label, and, to paraphrase Ludwig Wittgenstein, the grammar of our language is meant to model the structure of the universe. We think in words, and our words mirror the world. But for Einstein, thinking was never a linguistic endeavor. Language may be necessary for communication of our thoughts to others, but he thought it stood in the way of connecting ourselves to the world itself and to the world of our own mind. "For me it is not dubious that our thinking goes on for the most part without use of signs (words) and beyond that to a considerable degree unconsciously. For how otherwise should it happen that sometimes we 'wonder' quite spontaneously about experience? This 'wondering' seems to occur when an experience comes into conflict with a world of concepts which is already sufficiently fixed within us."[5] Words and concepts were not to be trusted, as they took us away from the real world and our real experiences within it.

We needed to return to our world of wonder, which puts us in touch with our inner images and not the abstract artificial concepts we use to categorize them. Images were one avenue to a direct link, and Einstein's science would always have a very visual, pictorial sense to it full of thought experiments, models, and metaphors. But another nonlinguistic way of connecting to the real elements of the universe is music. For Einstein, whether through performance or listening, music brought an experience of unification, connection, and order that did not make use of words or linguistic notions. We experience music directly, viscerally. It would not be exaggeration to say that Einstein thought music to be spiritual in the deepest sense.

But the elevation of music and mental pictures above the linguistic was a result not only of his love of them, but of his lifelong distrust of language, which probably stemmed in part from Einstein's difficulties as a child. These developmental problems have given rise to a number of well-known myths: Einstein was a late talker. Einstein was a poor student. Einstein's teachers thought he was stupid. Einstein failed mathematics. Einstein was autistic. The perpetuation of these stories derives no doubt from the icon of Einstein as a rebel, as someone who achieved greatness while not conforming to the received image of "the perfect child." When coupled with insecurities about our own perceived lacks, this mythological Einstein becomes useful. "It is okay if I fail at this task or do not achieve that goal," we tell ourselves. "After all, Einstein became the world's greatest scientist after failing math." Einstein the myth gives us hope in the face of failure. Unfortunately, Einstein the man is not identical to Einstein the myth, and while there is some truth to some of these stories, others are simply false.

Einstein did have issues with speech as a young child. He started speaking later than most children, but the degree to which this occurred is surely exaggerated. The story attributed to the great historian of ancient mathematics Otto Neugebauer, for example, is no doubt wrong. "It seems that when Einstein was a young boy he was a late talker and naturally his parents were worried. Finally, one day at supper, he broke into speech with the words, 'Die Suppe ist zu heiss.' (The soup is too hot.) His parents were greatly relieved, but asked him why he had not spoken up to that time. The answer came back:

'Bisher war Alles in Ordnung.' (Until now everything was in order.)"[6] It is a cute story, but there is good reason to think that Einstein was talking well before then. We have, for example, a letter from his grandparents praising the two-year-old Einstein for his "droll ideas," and while we cannot be certain that his conveying of these ideas was linguistic, it would seem unlikely if it were otherwise without note.

Similarly, there is no reason to suspect that Einstein had autism. While some, such as Temple Grandin[7] and Simon Baron-Cohen,[8] have argued that certain biographical facts about Einstein point to traits that are consistent with a diagnosis of autism, there are a number of other facts that support the opposite conclusion. Although it is certainly true that Einstein's delayed speech, social awkwardness, lack of concern for socially prescribed dress, and intelligence provide reason to suspect that he may have been somewhere on the spectrum, he also was capable of deep and recurring emotional bonds with others, had a lively sense of humor, and placed empathy for human suffering at the core of living a meaningful human life. Is it possible that Einstein had Asperger's? Of course, but there is not convincing evidence to support the postmortem diagnosis.

We do have evidence, however, to dispel the claim that Einstein failed mathematics in school or that he was not accomplished at mathematics as a child. From his report cards to glowing mentions in his mother's correspondences, we know that Einstein's grades, especially in math and science, were almost always at the top of his class. Einstein himself reports his extensive independent forays into the realm of mathematical thought during high school. "At the age of 12–16 I familiarized myself with the elements of mathematics together with the principles of differential and integral calculus. In doing so I had the good fortune of hitting on books which were not too particular in their logical rigor, but which made up for this by permitting the main thoughts to stand out clearly and synoptically. This occupation was, on the whole, truly fascinating; climaxes were reached whose impression could easily compete with that of elementary geometry—the basic idea of analytical geometry, the infinite series, the concepts of differential and integral."[9] These reflections, of course, concern Einstein's ability to teach himself mathematics and not his engagement in the classroom. But while we do have grades to show that the interest carried over to his formal instruction, it is also the case that Einstein disliked high school despite his success.

In 1889, at age nine, Einstein moved from Petersschule to the Luitpold Gymnasium, where he was taught not only math and science, but ancient history, art, literature, and classical language. One might think that such well-roundedness would have appealed to Einstein, but the content for the most part was tainted by the school's pedagogical approach. The teaching stressed discipline over free thought, rote memorization over understanding, and intellectual authority over critical questioning. Einstein famously compared the educational approach of the German high schools to the militaristic mindset that was rampant in Germany at the time. "The teachers in my elementary school appeared to me like sergeants, and the gymnasium teachers like lieutenants."[10]

Philipp Frank, a friend and colleague of Einstein's, explains this quotation in a fuller context:

> The sergeants in the German army of Wilhelm II were notorious for their coarse and often brutal behavior toward the common soldiers, and it was well-known that with the troops completely at their mercy, sadistic instincts developed in them. The lieutenants, on the other hand, being members of the upper class, did not come into direct contact with the men but they exerted their desire for power in an indirect manner. Thus when Einstein compared his teachers to lieutenants and sergeants, he regarded their tasks to be the inculcation of a certain body of knowledge and the enforc-

> ing of mechanical order upon the students. The pupils did not view the teachers as older, more experienced friends who could be of assistance to them in dealing with various problems of life, but rather as superiors whom they feared and tried to predispose favorably to themselves by behaving as submissively as possible.[11]

The Luitpold Gymnasium was significantly different from Petersschule in one respect: Einstein was not the only Jew. Indeed, there were enough Jews attending that his religious education requirement could be satisfied with a class in Judaism for the Jewish students, taught by pillars of the local Jewish community. While his religion classes fascinated him in elementary school, the same cannot be said for high school. In response to a letter honoring his fiftieth birthday, Einstein wrote to Heinrich Friedman, one of his former Jewish studies teachers, "I was deeply moved and delighted by your congratulations. How vividly do I remember those days of my youth in Munich and how deeply do I regret not having been more diligent in studying the language and literature of our fathers. I read the Bible quite often, but the original text remains inaccessible to me. It certainly was not your fault; you have fought valiantly and energetically against laziness and all kinds of naughtiness."[12] The last sentence not only betrays the lack of seriousness with which Einstein pursued his religious learning in high school, but also may be of great comfort to Hebrew school teachers everywhere that theirs are long-standing struggles.

Einstein's lack of interest in studying Hebrew and other Judaic matters was the result of a major shift in Einstein's worldview. The deep religiosity of his earlier years disappeared, replaced by a thoroughgoing skepticism for authorities of all types, including religious, and with a picture of a wholly material universe guided by rational principles discoverable through scientific investigation. While the first of these factors is usually chalked up to a combination of a rebellious personality coupled with a reaction to both the schooling and the militaristic mindset of the time in Germany, the second is traced back by Einstein to two events in his life.

The first of these occurred when he was quite young. "A wonder of such nature I experienced as a child of 4 or 5 years, when my father showed me a compass. That this needle behaved in such a determined way did not at all fit into the nature of events, which could find a place in the unconscious world of concepts (effect connected with direct 'touch'). I can still remember—or at least believe I can remember—that this experience made a deep and lasting impression upon me. Something deeply hidden had to be behind things. What man sees before him from infancy causes no reaction of this kind; he is not surprised over the falling of bodies, concerning wind and rain, nor concerning the moon or about the fact that the moon does not fall down, nor concerning the differences between living and non-living matter."[13] The fact that the needle of the compass always points north shows that it is subject to a force. But unlike a string, the thing pulling it is not visible and cannot itself be directly experienced. Yet, it is not capricious. The behavior of the compass is completely determined by a simple and understandable rule. The universe, then, must have a set of such rules governing it, rules that describe the working of the invisible forces that affect the visible world.

But it would be the second event that shaped Einstein's belief about how one comes to know these rules. "At the age of 12 I experienced a second wonder of a totally different nature: in a little book dealing with Euclidean plane geometry, which came into my hands at the beginning of a school year. Here were assertions, as for example the intersection of the three altitudes of a triangle in one point, which—though by no means evident—could nevertheless be proved with such certainty that any doubt appeared to be out of the question. This lucidity and certainty made an indescribable impression upon me."[14] Einstein had to study Euclidean geometry, and, unlike its effect on most of us, it changed his life. Euclidean geometry begins with definitions, axioms, and postulates. The definitions explain what is meant by basic geometric terms. The axioms are simple mathematical statements that are not peculiar to geometry—for example, equals added to equals yield equals. The postulates are particularly geometric statements that are so straightforward as to be seemingly beyond any possible doubt—"two points define a line," "three points not on the same line define a plane," and "around any point you can draw a circle of any size"—and clearly they were so to the young Einstein. If one takes these undeniably true basic building blocks and combines them using deductive logic, the results are scores of theorems—that is, other truths whose justifications rest only on the basic undeniable starting claims. Combining this insight with the picture of the universe derived from the compass experience, we get a worldview in which there are invisible forces guiding the universe according to fixed rules which can be discovered and comprehended through human reasoning.

The form of this reasoning was provided to Einstein by the reading of Talmud—not the collected thoughts of the ancient rabbis, but rather Max Talmud, a medical student who would dine on Thursday evenings with the Einsteins. A standard custom among Jewish families at the time was to have a student over for a weekly meal, and Max Talmud, who would later change his name to Talmey, was a regular guest. Wanting to show his gratitude, but being an impoverished student, Talmey thought that he could curry favor with Hermann and Pauline if he enthusiastically engaged their intellectually excitable son. Inexpensive popular science books were a trend in publishing at the time, and so on his weekly visits Talmey would often bring one to Albert and the two would enthusiastically discuss the contents. Of these, Aaron David Bernstein's multivolume *People's Books on Natural Science*, which opens with a discussion of the speed of light and goes on to cover topics from stellar astronomy to nutrition, was so influential that it warranted a mention by Einstein in the remembrance of his intellectual development he wrote at age seventy.

Between the books on geometry and the books on science, Einstein's picture of the universe radically changed with the end of his youthful religious period. "I came—despite the fact that I was the son of entirely irreligious (Jewish) parents—to a deep religiosity, which, however, found an abrupt ending at the age of 12. Through the reading of popular scientific books I soon reached the conviction that much in the stories in the Bible could not be true. The consequence was a positively fanatic [orgy of] freethinking coupled with the impression that youth is intentionally being deceived by the state through lies; it was a crushing impression. Suspicion against every kind of authority grew out of this experience, a skeptical attitude towards the convictions which were alive in any specific social environment."[15] Einstein's rejection of his religious beliefs was coupled with a rejection of the proclamations of the government and would lead to one of the bolder actions of Einstein's youth.

When he was fifteen, his father and uncle's electronics company failed. They had lost a major contract in Munich, thereby sinking the business, but Hermann and Jakob believed that the manufacture of electrical components was the right business at the right time, and that the problem was location. Once again reaching out to Pauline's wealthy relatives, they received funding to shutter the factory in Munich and relocate in Italy, first Milan, then Pavia, where the prospect of electric lights appeared brighter. The family moved but left Albert behind to finish his final years at Luitpold Gymnasium.

Graduation from a gymnasium was required for matriculation at a German university, and this was the family's plan for him—university, then a respectable profession. The long view, however, was not the adolescent Einstein's focus. Now living with a family who took in boarders, he felt alone. All he had in life was school, which he hated. This sentiment affected his attitude and behavior in the classroom. While it is not clear whether it was Einstein himself or school officials who initiated the break, after six months Einstein left Luitpold Gymnasium. According to one telling, he was encouraged to leave if not outright expelled. "Your presence in the class is disruptive and affects the other students,"[16] the letter from the school read. By another telling it was not formal expulsion but rather Einstein who chose to leave, getting a note from a doctor—Max Talmey's brother—which claimed that Albert was suffering from a nervous condition and should be excused from his studies.[17]

But whether he voluntarily left or was booted out of school, Einstein was weighing an additional factor. All German boys were required to serve in the military unless they left the country before age seventeen. He was then sixteen, and his family's relocation to Italy was the perfect reason to leave, so he traveled to Milan to join them.

He loved everything about Italy: the art, the climate, the people. He saw a different approach to life, one that suited him. Losing the sense of cultural oppression that led him to be dark and moody in Germany, the adolescent Einstein shone. "A freer life and independent work made of the quiet, dreamy boy a happy outgoing, universally liked young man,"[18] wrote his sister, who spent much time with him in this period.

But something had to be done about his education, because without the gymnasium diploma the German universities were not an option. The solution was to look outside of Germany. He needed the instruction to be in German, so Switzerland seemed the natural choice, and given his love and aptitude for mathematics the Swiss Federal Institute of Technology, the Eidgenössische Technische Hochschule, or ETH, seemed to be the place for Einstein, who wanted to study mathematics or (if his parents had their way) electrical engineering. Best of all, entrance required no diploma, just passing an entrance exam. The plan was to study on his own, take the exam, and register the following year.

The prospect of self-education—learning without authority —pleased Einstein. He continued to read in physics, philosophy, and classical German literature, but because he allowed his personal interest to guide him Einstein did not acquire as well-rounded an education as he would have received from formal schooling. Einstein took the entrance exam, but contrary to plan he did not pass. While his scores on the mathematical portions were well above standard, other areas lacked and so he could not be admitted.

An administrator, seeing the promise in the young man, suggested a school nearby in the Swiss village of Aarau. Completing his education there would fill the gaps, and, with a Swiss diploma, admission would be assured.

The prospect of returning to the classroom did not thrill Einstein. He had left the first time for a reason, and here he would once again be away from his family with nothing but formal schooling. Fortunately, both the pedagogy and the living arrangements would be different. At the Aargau Kantonsschule, independent and critical thought was valued. Einstein was free to express himself in his playfully impudent fashion, the teachers seeing themselves as fellow travelers and guides on the path to knowledge, not authorities to be obeyed. Contact with the world was seen as valuable in learning, and Einstein spent time thinking and conversing on hikes the teachers would lead. He loved the school and became a much more social person than he had ever been in Germany. His grades were quite good, receiving top marks in mathematics, physics, singing, and violin, and high marks in history, biology, drawing, and chemistry (for which he had to receive individual tutoring because of his lack of previous exposure). His lowest grade was in French—in both diligence and achievement. By request he was excused from gymnastics, and as a foreign-born student he was exempted from military instruction.

His allergy to all things martial led him to avoid not only instruction in Switzerland, but also service in Germany. When he left to join his parents in Italy, he surrendered his passport as a symbolic relinquishing of his German citizenship. But he wanted to act in more than a mere gesture, and once he had moved to Switzerland—and after pestering his father for quite a while—official notice was sent that Einstein was renouncing his status as a German national. He would be stateless, a citizen of the world.

In Aarau he lived with the Winteler family. The father, Jost Winteler, was a teacher at the school, and he and his wife, Rosa, would become second parents to Einstein. When the conversation around the dinner table was not about academics, it was about current events and politics, the Wintelers' liberal, pacifistic leanings matching Einstein's. He loved the new school, he loved the new family, but most of all he loved the Wintelers' daughter Marie, who became his first sweetheart. While the young romance would ultimately go nowhere, he never lost his sense of family with the Wintelers—indeed they literally would eventually become family when their son Paul married Einstein's sister, Maja.

His love for institutional learning temporarily rekindled, in 1896 he left the sleepy village of Aarau and entered the ETH to study mathematics and physics, planning on becoming a high school teacher—a position that afforded respect and social standing. Zurich would be his home for the next four years. The Zurich of Einstein's college days was a cosmopolitan playground filled with young people from across the Continent. Radical new ideas were in the air, and a sense of openness abounded. Einstein reveled in this environment, spending time in the cafés discussing science and politics. He took up sailing and frequently played duets, his violin eagerly accompanying the piano of various attractive young ladies around town.

He studied mathematics, philosophy, and physics—both experimental and theoretical—as he pursued the track in science, something somewhat unusual at the ETH, where the students were almost universally studying engineering. Indeed, in Einstein's class, there were only four others majoring in physics. Three were Swiss: Marcel Grossmann, Jakob Ehrat, and Louis Kollros. The other was Serbian, but that was not the most unusual thing about her. Mileva Maric was smart, sarcastic, and strong-willed. It was highly unusual for a woman to pursue education at the collegiate level at the time; indeed, Switzerland was one of the few places where it was even possible for a woman to enroll in a university. The move was not popular with her family, but she was determined nevertheless not only to earn a degree, but to study the difficult but burgeoning field of physics. Undertaking such a bold path and reaching this point demonstrated her truly remarkable spirit.

Einstein took to his new colleagues and enjoyed the time he spent with them, but because he preferred coffeehouses to lecture halls, that time was spent less frequently in the classroom than it should have been. Rather than following the prescribed course of study, Einstein thought that he would be better served reading on his own about the latest advances and developments in physics and discussing them with his peers. He was at a university to learn physics, and to him physics is what physicists were doing now, not some set of dead and settled beliefs, mere recipes for calculation set out in sterile textbooks. The classes, he thought, were conceived all wrong. They were far too conservative in scope to be worth his time.

This approach was that of Heinrich Weber, who led the physics department and taught the course in theoretical physics. Weber was a student of Hermann von Helmholtz's, the greatest name in nineteenth-century German science. Helmholtz, a physicist, physiologist, mathematician, and philosopher, worked on building a single coherent system that accounted for mind and world and the relation between them. He tried to account for questions of space, perception, weather, non-Euclidean geometry, and human knowledge in a framework that sprang from Newtonian physics. Weber took this project seriously, and because in his mind physics ended with Helmholtz, his course on theoretical physics was designed as a history of the field leading up to and ending with Helmholtz. These lectures were famous for Weber's excellent presentation style and clarity. Even Einstein praised his teaching at first: "Weber lectured masterfully on heat (temperature, heat quantities, thermal motion, dynamic theory of gasses). I eagerly anticipate every class of his."[19] But Weber's limited horizon directly challenged Einstein's notion of the discipline, and Einstein's eager anticipation for Weber's classes turned to annoyance with Weber's lack of enthusiasm for the cutting-edge theories.

If Weber would not give him what he thought he needed, then he would get it on his own. Einstein turned away from classes and went back to self-study. The problem, of course, is that he needed to pass his exams. Class notes were thereby essential, but not being in class made this difficult. Fortunately for Einstein, his friend Marcel Grossmann not only was in class, but took incredibly clear and detailed notes, and he was willing to share them with Einstein. It was the first—but certainly not the last—time Grossmann bailed Einstein out of a tough situation that he created for himself out of arrogance. Einstein studied hard with Grossmann, and as a result they finished first and second in their class.

Einstein's success reinforced the idea that he always knew better. Although Einstein worked hard on anything that interested him, his lack of respect for the faculty and the curriculum led to a falling out with Weber, who dismissed Einstein as impudent and unserious.

Things were even worse with another of his physics professors, Jean Pernet. While Einstein had his issues with Weber, he always worked hard in Weber's lab. Weber was not only well known and well respected, he was well connected with the elite of the time and had one of the largest and best-equipped setups in Europe thanks to personal support from electrical magnate Werner von Siemens. This made Weber's laboratory a playground for Einstein's curiosity, and his dedication earned him top marks. The same, however, cannot be said for Einstein's work in Pernet's lab. Einstein treated Pernet with far less than the formality expected by a professor. He spoke irreverently and would make a show of crumpling the laboratory instructions into a ball and tossing it into the wastepaper basket before beginning the experiment as he saw fit. This behavior hurt Einstein, both academically and literally. Pernet, upset with Einstein's attitude and poor attendance, had him formally reprimanded and gave him a failing grade for the class. As to bodily injury, one day while ignoring the work he was supposed to do, he instead decided to see if he could evacuate a glass jar to the point where light could not travel through it. If light travels as a wave, then something must be doing the waving. Physicists called this medium the light-bearing, or luminiferous, aether. If the classical theory was correct, then pumping everything out of the jar—not only the air it contains but all the aether as well—should make it impossible for light to pass through. Einstein, testing whether the pump would cause not only an air vacuum but an aether vacuum as well, ran the suction pump too long. The lack of pressure caused the jar to implode, and the flying glass shards badly injured Einstein's hand. Einstein would have to wait to do additional cutting-edge research.

Although often described as a loner, Einstein did actively

seek the company of others, especially women. When he started high school, he and the Wintelers' daughter, Marie, began an extended courtship that lasted into Einstein's time at university, but Einstein ended it abruptly when his romantic interest turned to another woman, his classmate Mileva. In Mileva he saw a strong, smart woman whose personality and passion for physics matched his own. Mileva had no interest in a traditional domestic life. She was intelligent, driven, and destined for something greater, no matter what the culture would expect. Einstein admired her ability, and when the two were apart during summers and during Mileva's semester studying in Heidelberg, he often wrote to her of his plans for the two of them—Dolly and Johnny, as they affectionately referred to one another—to be a scientific team, facing down the great problems of the universe together.

The two became constant companions. Even though they had separate residences, Einstein frequently referred to "our household," giving reason to believe that they were, in fact, virtually living together. Referring to himself in a letter to her as "your good colleague and fellow coffee-guzzler," Einstein had drawn Mileva into his approach to learning, self-study, avoiding classes, and thinking about topics the professors would rather they not consider. While Einstein could make up for this with intensive study and Grossmann's notes at the end of the year, Mileva could not. In their second year, Einstein finished at the top of the class, but Mileva had to postpone her examination. In the fourth year, Einstein's luck ran thin and he barely passed his exams, scoring second to last. Mileva did even worse, failing and thereby not graduating. She would have to study on her own for another year, a single retake being allowed. She had to pass next time or she would never graduate.

His family knew of Mileva, and at first they were civil about the relationship. Einstein wrote that when his mother—his "old lady," as he referred to her in the letters—found Mileva's picture, she studied it carefully and afterward he "had to endure much teasing."[20] He frequently sent regards from his mother and sister, and at the bottom of one letter Maja wrote, "I would have liked to write you too, but A. wouldn't let me"[21]—typical big brother.

But as things got more serious between them, his family soured on the relationship. Cohabitation might be common today, but at the time they were pushing the bounds of social acceptability. He was still young, she was older, and the Einsteins always had in mind a stable upper-middle-class life for their son, one in which you first get a good job and then settle down with someone respectable. They constantly compared her to the scorned Marie Winteler, whom they adored and whose family they would have loved to have as in-laws.

When news of Mileva's failed exams arrived, the situation came to a head. "So we arrive home, and I go into Mama's room (only the two of us). First I must tell her about the exam, and she asks me quite innocently 'So, what will become of your Dollie now?' 'My wife,' I said just as innocently, prepared for the proper 'scene' that immediately followed. Mama threw herself on the bed, buried her head in the pillow, and wept like a child. After regaining her composure she immediately shifted to a desperate attack: 'You are ruining your future and destroying your opportunities.' 'No decent family will have her.' 'If she gets pregnant you'll really be in a mess.' With this last outburst, which was preceded by many others, I finally lost my patience. I vehemently denied that we had been living in sin and scolded her roundly."[22] His mother's warnings went unheeded.

The next year was difficult. The expected step for Einstein was to become an assistant to a professor of physics somewhere. Science at the time worked on something akin to an apprenticeship model. Graduates would be employed in the lab of a scientist, doing the grunt work and learning the ways of research—all the while writing their dissertation on their own, which would then be submitted to a university faculty. Einstein wrote letters of inquiry to physicist after physicist, eventually writing to every single one on the continent of Europe without receiving a single positive reply. He was shut out of the scientific community. Because job applications require a checking of references, any scientist who received a letter from a graduate of the ETH would immediately contact one of the physics professors there to inquire about the young man's qualifications. Einstein's arrogance, lack of diligence, and unwillingness to follow directions certainly would not have led a prospective employer to conclude that the recent graduate was well suited to be his assistant. With each rejection, Einstein became more and more deflated.

He needed money, so he found a job as a substitute teacher for six months and then another position tutoring the young sons of a gymnasium teacher. Einstein enjoyed the teaching experience and worked hard to find new and innovative ways to get the boys to learn. So confident was he in the superiority of his approach to the formal education they were receiving that he demanded the boys' complete education be turned over to him entirely. Needless to say, their father—a teacher—did not think much of this ultimatum, and the tutor was dismissed.

A permanent teaching position seemed attractive, but lacking Swiss citizenship he knew such jobs were not open to him. So Einstein saved money and went through the process to become officially Swiss. He was no longer a man without a country, but despite his newfound status he remained a man without a job. Old concerns reemerged, and as a Swiss citizen he was once again expected to perform compulsory military service. He went for his physical, and to his great relief he failed—flat feet and varicose veins kept him out of the army.

He refocused on professional science and in 1900 published his first paper, "Deductions from the Phenomena of Capillarity." It appeared in the top physics journal of the time,

Annalen der Physik. Einstein hoped it would pave the way for him, but no response came.

In 1901 Einstein was searching desperately for work while living with his family in Italy. His future was uncertain, and Mileva, back in Zurich studying for her exam retake, found her future even more indefinite. They decided they needed some time alone. Einstein convinced her to meet him for a long romantic weekend at Lake Como. "I just don't let up! You absolutely must come to see me in Como, you sweet little witch. It will cost very little of your time and will be a heavenly joy for me. We'll be back in three days."[23] They had a delightful time, hiking and exploring the town and gardens while lodging at a small inn. But Einstein's mother's fear came to pass, and Mileva became pregnant.

This element of Einstein's life was hidden from biographers until 1986, when letters between Einstein and Mileva were discovered. Mileva returned home to Serbia and gave birth to a daughter, Lieserl. In his letters Einstein asks about her, expresses his love for her, and shows concern for Mileva, whose delivery was complicated. But beyond this correspondence, Einstein never mentioned her existence to his family or friends.

The pregnancy added tremendous pressure for Einstein to find a job. If he was going to be a father, he needed a stable income. He wrote that he would look for any job anywhere, whether it was teaching or selling insurance. Whenever Einstein was in dire straits, it seemed his friend Marcel Grossmann always came to his rescue. When exams loomed, Grossmann's notes and tutoring got him through college. Now, when Einstein most needed a job, he again turned to Grossmann. His friend consulted with his father, who through connections was able to get Einstein hired in the patent office. After a period of waiting for the position to move through the bureaucratic process, an opening appeared for a patent clerk with a scientific background, and Einstein was hired in Bern.

But this stroke of luck in obtaining a good government job presented a new challenge that could cost him the position if it were to become known that he was to have a child out of wedlock. Although he found a flat that would be family-friendly, a pregnant Mileva, much less a Mileva with Lieserl, could not join him.

We do not know exactly what happened to Lieserl. The custom at that time was for the children of unmarried parents to be adopted, usually by a family member or a close friend. This seems to have occurred, as news of Lieserl continued in correspondence for a little while. Mileva moved back to Zurich, where she received word that Lieserl had contracted scarlet fever. We do not know whether she survived; there are conflicting stories, some in which she died and others in which she lived, blinded by the disease, to adulthood with an adoptive family. Most likely she passed away, but we do know that Einstein never met his daughter.

The situation took its toll on Mileva. The physical and emotional stress of the pregnancy and her recovery from the birth was considerable and took her away from her studies. She once again failed her exams. But now, without a child, at least she and Einstein could marry. She would not become the pioneering female physicist she had envisioned, instead finding herself trapped in the bourgeois role of wife and mother that she worked so hard to avoid. Her mood became increasingly dark.

Her emotional state was not helped by Einstein, who came to view her less and less as a scientific partner. He had established himself at the patent office and came to find the work enjoyable and challenging. His job was to take patent applications and figure out the mechanisms behind the claims and determine their originality. Given Einstein's interests, dealing with technology and creativity was agreeable. It also gave him time—if he worked hard enough to get a full day's work done early—to sneak in some work on his physics. If he wasn't carrying on his extracurricular efforts on his own at the office, then he was doing so with his friends Conrad Habicht, Maurice Solovine, and Michele Besso, who formed a reading group they called "the Olympia Academy," which considered questions of physics and philosophy. He had created an intellectual life for himself without Mileva.

One additional roadblock was Einstein's family, which had forbade the marriage. Einstein struggled with this edict until he was called to his father's deathbed. Hermann gave Albert his permission to marry Mileva, but he then asked his son and everyone else to leave so he could die alone.

Einstein and Mileva were married in a small ceremony without any family present, and with Habicht and Solovine as witnesses. Afterward came a small celebration and then Einstein was to bring his bride home and carry her across the threshold of their apartment to begin their lives as husband and wife—except that Einstein had locked his keys inside the apartment. His first act as husband was to awaken an unhappy landlady in the early hours of the morning to let them in.

It was an omen. The marriage began happily as the two enjoyed a brief honeymoon phase, but Einstein saw marriage as a form of convenience, expecting Mileva to take over the pedestrian duties of everyday life, leaving him free to think about physics. She happily did so for a while. But soon the novelty wore off and Mileva felt alienated and neglected. The more she voiced her displeasure and fell sullen, the more Einstein distanced himself from her. This in turn caused Mileva to grow even more resentful, and the marriage spiraled downward.

Things improved temporarily, however, when a few months later Mileva was once again pregnant, giving birth on May 14, 1904, to their son Hans Albert. Einstein was delighted, and the newborn child brought new energy to the relationship. But the old problems soon resurfaced. A full-time job, a dysfunctional marriage, and a newborn child could sap anyone's creative energy. This was Albert Einstein's life as the calendar turned to 1905. Working all day in an office with no access to the resources of a university library, Einstein was hardly enjoying the lifestyle of someone who would create an intellectual revolution that would reshape history.

2

The Miracle Year

In 1905 Albert Einstein, despite not working as a researcher, planned to transform physics. In a letter to Habicht, who had just moved away from Bern, Einstein refers to elements of his projects for the year as "very revolutionary." Anyone else uttering such words could well be considered audacious, bordering on naïve. Not only did Einstein seem highly unlikely to be the person best positioned to change the course of the oldest and most established field in science, but the field itself did not seem to need changing.

The philosopher Thomas Kuhn argued that scientific revolutions are preceded by a period of crisis in which anomalies force scientists to reconsider the basic concepts, methods, and presuppositions underlying standard practice.[1] Typical scientists, he contended, hate questioning the basis of what they do—they prefer to simply go and do it. So the period before a scientific revolution is very uncomfortable for those in the field who are compelled to think philosophically.

But this was not the case in 1905. There were a few problem areas and a couple of strange results that needed to be accounted for, but no one expected an imminent Copernican-style revolution. Newton's mechanics and James Clerk Maxwell's electrodynamics were entrenched and, by and large, working perfectly well. The boat was in calm waters, and rocking it seemed a waste of time and energy.

Einstein thought what he was about to do was revolutionary because he had a coherent vision of the universe that disagreed with the accepted view in crucial ways. When we understand matter, light, and space differently, he thought, everything would have to change. To make this happen, two things were needed: the details of his picture would have to be rigorously developed, and important people in the physics community would have to take his work seriously, the work of a professional nobody. He knew he needed incredibly strong arguments and irrefutable results, but he also knew that, no matter what he wrote, it would not be considered if it was the work of a mere patent clerk. Shut out of the academic physics community, he knew that if he could not work in the lab of an established physicist he would need to enter by a side door—and for that he would need a doctorate.

Today, if you want a Ph.D., you need to be accepted into a graduate program and satisfy a program of study that includes coursework, exams, and a dissertation process under an adviser resulting in research that expands the field. In Einstein's time, things were less formal and students submitted their work to university faculty, who then determined whether it was novel and important enough to merit the degree.

Einstein's first attempt at a Ph.D. came right after graduating from the ETH. Working in Weber's lab, he explored the thermoelectric Thomson effect.[2] When wires of different materials are connected and electrical current is sent through them, the wires heat up to different temperatures. This did not seem too odd, but more surprisingly, the effect works in the other direction—that is, by heating wires made of different materials joined together one can actually generate electricity. Exploring this connection between heat and electricity was classic Einstein, as he loved looking for insights in the relations between seemingly different subfields of physics. But in 1900, when he first began working toward a doctorate, the effort fell flat.

Einstein next struck out on his own and wrote a dissertation concerning the forces between atoms in a gas in 1901, submitting it to Professor Alfred Kleiner at the University of Zurich, across town from the ETH. No copies of this manuscript survive, but ultimately it was rejected by Kleiner. Some have contended that it was a matter of quality, as Einstein himself would later call it "worthless,"[3] while others have concluded that the problem was "Einstein's attack on the scientific establishment,"[4] especially Ludwig Boltzmann, a very esteemed colleague of Kleiner's. If true, the latter assessment would, in a sense, be ironic because it was Boltzmann's picture of the universe that was the basis of Einstein's revolutionary ideas.

In the preceding generation, a question about heat led to a concern about the nature of physical laws and the structure of the universe. Energy comes in a variety of forms, forms that can be transformed from one to another, but whenever such a change occurs the amount of energy we get out is never quite as much as the amount we put in, because some is always lost in the transfer. Consequently, efforts to create a perpetual motion machine always fail because to get something to move, we need to take some form of energy and convert it into kinetic energy —that is, the energy of motion. This conversion is always incomplete, we always lose some energy, so the machine slows down until eventually it stops. This regularity is expressed by physicists as a principle called the second law of thermodynamics, and is written out in terms of a quantity termed "entropy."

Despite its title as a law, a major dispute arose about the meaning of the second law of thermodynamics. Some, such as Rudolf Clausius and the young Max Planck, argued that as a law it must be both universal and true; that is, it must apply always and everywhere. The second law of thermodynamics says that entropy increases in a closed system (a system where energy is not added), so if the second law of thermodynamics is a real law, entropy necessarily increases. Arguing otherwise were scientists in line with Boltzmann, who contended that we cannot just state these laws as numerical correlations without having a sense of the operative physical factors bringing them about. In other words, we need to pop the hood and see what is happening underneath. Boltzmann was an adherent of the kinetic theory of gases, wherein matter comprises molecules, and heat is the energy of their motion. The second law needed to be worked out in terms of the interactions of these particles.

Many important and respected scientists before the turn of the twentieth century did not accept the atomic view of matter. Atoms and molecules were unobservable, and this group believed it was wrong to try to make sense of empirical results in terms of nonempirical entities—if we cannot see atoms, why would we base our understanding of physics and chemistry on them? But Boltzmann did, and he translated entropy talk into ideas about the statistical distribution of the speeds of the molecules. But when we talk about large-scale statistical relations, unexpected situations must be expected to pop up occasionally. Everyone who buys a lottery ticket should expect to lose, but someone does eventually win. The thermodynamic version of the lottery implies that when we talk about entropy, we should expect at any given time it will increase, but there will be rare cases where on its own it decreases just by luck of the draw. The second law of thermodynamics, Boltzmann held, was a statistical generalization that says that entropy tends to increase.

Einstein shared Boltzmann's picture of a universe full of atoms bouncing around, their behavior governed by statistically derived results. While the details may not have been exactly Boltzmann's, Einstein's second attempt at writing an acceptable dissertation worked along these conceptual lines.

Typical of Einstein, he sent his dissertation to Kleiner, absolutely certain that it would be accepted. Einstein waited and waited, but no word about his dissertation came. Had Kleiner rejected his effort out of hand and not written back? Did he not understand it? The fact was that, several months later, Kleiner still had not gotten around to reading it. So, after expressing some rude words about the man in letters to his closest friends, Einstein went from Bonn to Zurich to see Kleiner, who received him warmly. He wrote to Mileva, "I spent all afternoon at Kleiner's in Zurich telling him my ideas about the electrodynamics of moving bodies, and we talked about all sorts of other physics problems. He's not quite as stupid as I'd thought, and moreover, he's a good fellow. He said I could count on him for a recommendation anytime. Isn't that nice of him? He has to be away during the vacation and hasn't read my paper yet. I told him to take his time, that it's not pressing. He advised me to publish my ideas on the electromagnetic theory of light of moving bodies along with the experimental method."[5] After the visit, Kleiner did eventually make time to read it. He did not like what he read and encouraged Einstein to retract it. Einstein did and was able to recover the 230 franc fee that he had to pay as part of the dissertation submission process, but as a result he became quite bitter.

For a while, Einstein put aside the idea of getting a Ph.D., but then he realized that it would be both useful if he stayed at the patent office and essential if he was to continue pursuing an academic post. Given Kleiner's advice to publish his work on the electrodynamics of moving bodies—that is, the basis for the special theory of relativity—he sent in his work on the question as his third attempt. It, too, was rejected. Recognizing that something safe, conventional, and based on experimentation would be best, Einstein decided to send in part of a larger project, something that was practical and uncontroversial.

At the heart of Einstein's worldview in 1905 were atoms. With his friends in the Olympia Academy, he had read Ernst Mach, the German physicist and philosopher who championed a view called "positivism," in which only that which is observable can be said to be real. Mach used this view to argue that some of the most central elements of physics were, in fact, not real. These included Newton's absolute space, and atoms. Einstein was fascinated by Mach but thought he had to be wrong about atoms. What the world needed was overwhelming experimental evidence, but for this Einstein would need to first build the mathematical foundation. That groundwork was the topic of his dissertation.

The work of physics is to take a situation in the world, figure out which among the infinite elements of the system that could be observed are important, determine how to turn these elements into measurable quantities, and finally set out relations among these quantities whose truth can be experimentally tested. Physicists create equations—that is, mathematical sentences in which the arithmetic combination of terms on the left side of the equal sign yields the same number as the arithmetic combination of the terms on the right side. Those terms either are directly measurable themselves or could be calculated from other measurable quantities using other equations.

In Einstein's work in 1905, we see just such a recurring approach. Einstein's style was to consider a system that can be described in two different ways and figure out how those ways relate to each other. Einstein was a synthesizer; he liked to bring together accounts that seemed to be contrasting takes on the same situation and then argue that they were just different ways of seeing the same thing. Since the accounts are not competing but equivalent descriptions, we can set them equal to each other and find new relations, thereby producing understanding and insight into the nature of that system.

Einstein did just this in his dissertation "A New Determination of Molecular Dimensions." Atoms might not be observable, but this does not discount the existence of macroscopic phenomena that would allow us to discover aspects of their microscopic existence. Einstein's dissertation and the paper that would follow were a two-step process to providing irrefutable evidence that matter is made up of discrete building blocks.

Before Einstein, atomists focused on the behavior of gases. The measurable quantities of volume, pressure, and temperature had long been related to each other, and in a series of famous papers in the mid-nineteenth century James Clerk Maxwell figured out how to think of them in terms of large collections of microscopic molecules. If gases were really just a bunch of disconnected molecules in motion, then pressure would be the result of the particles bouncing off the wall of the container in which they were held. Since heat adds energy, higher temperature could be thought of as increasing the average speed of the molecules. Maxwell had made a few assumptions about the molecules to make calculation easier—they are spherical and infinitely hard and interact only by contact (in other words, we could think of a gas as if it were a cloud of billiard balls)—and he then derived what we call the "Ideal Gas Law," $PV = nRT$, in which P is the pressure, V is the volume, n is the number of molecules, T is the temperature, and R is the Rydberg constant, a number that makes the units come out correctly. This law makes sense to us intuitively. Blow into a balloon and it expands—meaning that if we allow the pressure on the outside of the balloon and the temperature inside to remain constant, the volume of the balloon increases if we add air molecules. If we blow into the balloon while holding it so that it cannot expand, then the pressure on the walls increases. If we tie the balloon off, thereby fixing the amount of air inside, and put it in the freezer, the balloon shrinks because its volume decreases with the temperature. Maxwell showed that all of this would be expected if we thought of gases as if they were made up of molecules.

Einstein wanted to remove the inference of the "as if." Molecules are not merely a heuristic device, a pictorial tool we use; they are real. To move from "as if" to "are," we need to strengthen the argument. Einstein did this by doing for liquids what Maxwell did for gases by considering dissolved substances in solution. Take a cup of tea and add a teaspoon of sugar. The sugar melts from solid crystals to a viscous liquid that distributes itself through the tea so that after a while every sip has roughly the same sweetness. Liquid flows, so it seems that it is not a set of solid, discrete molecules but rather is something smooth. Einstein set out to show that this smooth liquid really was made up of individual bits by considering how the liquid sugar distributes itself throughout the teacup.

To do this, Einstein makes assumptions similar to Maxwell's —that is, assume that (1) the sugar molecules are spherical, and (2) the sugar molecules are so much larger than the molecules of the liquid that from the perspective of larger sugar molecules the tea molecules can be seen as smoothly surrounding them.

Einstein considers two different ways of looking at the situation in the teacup. The sugar, he argues, would be both pushed and pulled. On the one hand, the melted sugar would be pushed forward by diffusion. All systems seek equilibrium, meaning that over time things become evenly distributed. The sugar would move from the small area it initially occupied to being randomly distributed throughout the teacup. On the other hand, the sugar would be pulled back by the viscosity, or thickness, of the liquid. The gooier the liquid, the slower the molecules move through it and the longer it takes them to spread out evenly. Einstein was able to work out equations for the push and pull in such a way that both were fully determinable by measurement except for two terms, one being the size of the molecule and the other being Avogadro's number, the quantity of molecules in a gram of the substance. Having two equations and two unknown terms, we can do some basic algebra and get values for them.

But notice what these unknown values are—the number of molecules and the size of molecules. These make sense only if there are molecules to have a size and a quantity. When we combine this work on liquids with Maxwell's similar work on gases, it seems that we have even more reason to believe in atoms as not just useful fictions, but real entities.

While the underlying desire is the theoretical goal of proving that matter is atomic, the dissertation itself is actually quite practical in that the equations can be used to tell us how two substances will blend. Whether one is mixing cement or blending flavoring into ice cream, Einstein's paper shows how to do it. As a result, his least famous work of 1905 is actually the most useful and thereby the most cited of all his papers.

Most important for Einstein, Kleiner was happy with it, although years later Einstein told biographer Carl Seelig that at first Kleiner was not happy enough, sending it back to Einstein because he thought the paper was too short. So, Einstein being Einstein, he claimed to have added a single sentence and sent it back.[6] This time it was accepted. Kleiner wrote, "The arguments and the calculations are among the most difficult in hydrodynamics and could be approached only by someone who possesses understanding and talent for the treatment of mathematical and physical problems, and it seems to me that Herr Einstein has provided evidence that he is capable of occupying himself successfully with scientific problems."[7] Einstein was now a doctor and could proceed to step two in the demonstration of atoms.

In his next paper of 1905, "On the Movement of Small Particles Suspended in Stationary Liquids Required by the Molecular-Kinetic Theory of Heat," Einstein takes the work he did in his dissertation and applies it to another case, what came to be called "Brownian motion."

Looking at liquids under a microscope, one discovers something odd. Small particles in the liquid can be seen to zigzag around randomly. Some thought that the motion was evidence that the particles had to be biological, that there were small animals moving about. But biologist Robert Brown showed that the phenomenon was replicable with clearly nonliving particles such as ground glass. So it had to be a physical phenomenon.

Some suggested it was caused by thermal currents in the liquid. Hot flows to cold, and maybe this was pulling the particles along. This was not the answer either, because neighboring particles should share similar motions, but they do not.

Others suggested an atomic explanation. Maybe it was an interaction of the atoms of the liquid striking the particulate matter. But the particles were too heavy. It would be like trying to move a bus by hitting it with balloons. The source of the Brownian motion was a mystery.

Einstein thought that the atomic explanation was on the right track, but that the phenomenon was more complicated than could be explained by one-on-one collisions—indeed, it is more complicated in exactly the same sort of way the sugar in tea example worked. The larger particles are surrounded by many tiny molecules bouncing off of them. But since the liquid around them is composed of molecules all moving in various directions, on average the molecular collisions from one side of the large particles would most likely be balanced out by the collisions from the molecules on the other side. Only if you had a significantly greater number of collisions from one side than from the other would enough force be generated to cause the large particles to move. If we take the molecular hypothesis seriously and think of the liquid in equilibrium—that is, the same energy of motion everywhere—then there should be no movement of the particles.

But the picture Einstein had of the microworld was much more chaotic. The molecules were not smoothly choreographed; rather, the temperature was an average, with some moving faster, others moving slower, and all moving in random directions. As such, the combined interactions of the molecules with the particles would be a result of chance. There was a possibility that a lot more molecular collisions could occur in one direction, but there was no reason to think it would happen in one direction rather than another. How could one mathematically model this situation?

Einstein realized that he recently had done just that. In his doctoral dissertation, he treated the molecules of dissolved sugar as larger spheres surrounded by a sea of liquid. That is exactly what was happening in this case. Einstein could use the same exact mathematical moves from his dissertation and apply them to Brownian motion.

> It will be shown in this paper that, according to the molecular-kinetic theory of heat, bodies of microscopically visible size suspended in liquids must, as a result of thermal molecular motions, perform motions of such magnitude that these motions can be detected by a microscope. It is possible that the motions to be discussed here are identical with the so-called "Brownian molecular motion"; however, the data available to me on the latter are so imprecise that I could not form a definite opinion on the matter.
>
> If it is really possible to observe the motion to be discussed here, along with the laws it is expected to obey, then classical thermodynamics can no longer be viewed as strictly

> valid even for microscopically distinguishable space, and an exact determination of the real size of atoms becomes possible. Conversely, if the prediction of this motion were to be proved wrong, this fact would provide a weighty argument against the molecular-kinetic conception of heat.[8]

Einstein's carefulness in the first paragraph, hedging as he does on whether the motion he is predicting is, in fact, Brownian motion, is to some extent legitimate in that his lack of access to university resources did leave him without the tools needed for the stronger claim. At the same time, there is little doubt in Einstein's mind that what he is describing is Brownian motion. This makes the challenge in the second paragraph something of a sucker's bet. Einstein sets out observation of the motion that he "predicts" will occur as something philosophers of science call a "crucial experiment"—that is, a result that if observed will justify a theory, and if not observed will falsify it. It is all or nothing, a sudden-death overtime for scientific theories. Einstein seems to be playing it honestly—go check and see if this motion is observable; if it isn't, then the molecular picture will have been defeated and we all need to stop believing in the reality of atoms. Of course, he makes this offer only because he knows that the motion has already been repeatedly observed.

But the wager requires more than generalities, and Einstein ends the paper by throwing down the gauntlet for experimenters. He considers particles 0.00004 inches in diameter suspended in water (he needs to specify the liquid so that the viscosity is known), and predicts that at 62.5 degrees Fahrenheit the particle would move horizontally (thereby neglecting the effects of gravity) about 0.000003 inches per minute. Since this is a displacement that was within the possibility of being observed in the lab, though not easily, Einstein concludes with: "Let us hope that a researcher will soon succeed in solving the problems posed here, which is of such importance in the theory of heat!"[9] The use of exclamation points is not common in scientific papers, and it shows both Einstein's sense of the importance of his work and his lack of concern with the usual marks of professionalism.

Thankfully, researchers did take up this challenge, and the French master experimenter Jean Perrin verified Einstein's calculation, earning the Nobel Prize in 1926. Perrin, with Einstein, was a strong proponent of the atomic theory and, also in line with Einstein, a prolific calculator of Avogadro's number. They both spent much time collecting radically different sorts of phenomena from across the spectrum of subfields in physics, all of which could be used to calculate Avogadro's number. Since Avogadro's number is the number of atoms in a fixed amount of a substance, each approach must give independent and additional support for the existence of atoms. After all, you cannot have a number of things without having the things to be numbered. But even after Perrin's work, which once and for all convinced even the most dogmatic holdouts in the scientific community of the existence of atoms, the two continued to seek out novel and disparate places to find Avogadro's number hiding in the equations of the disparate subfields of physics and chemistry.

Demonstrating the existence of atoms would be a major accomplishment for any scientist, much less one so junior, but this effort is widely regarded as the least important of Einstein's work in 1905. In addition to the structure of matter, he also considered a radical revision of our understanding of the nature of light.

In the nineteenth century, light was considered a wave for both experimental and theoretical reasons. Despite Newton's contention that light was a particle, the French physicist and engineer Augustin-Jean Fresnel and the British researcher Thomas Young produced optical effects that seemed to be explainable only if we consider light to be a wave. Waves add and subtract in particular ways when they flow through each other—"interfere," in technical terms—and there are numerous phenomena where this sort of interference can be seen. In these cases, light behaves as a wave.

Maxwell's equations, the four laws that govern electromagnetism, are named for James Clerk Maxwell, who discovered none of them and in fact made just one change to one term in one equation. But when Maxwell took these individual principles and put them together into a coherent, unified theory, they became incredibly powerful. One of the unexpected results was that these equations, which describe the behavior of electricity and magnetism, could also be used to describe the behavior of light. Maxwell's equations give us a description that has the form of a wave equation, so given the observable evidence light had to be thought of as a wave of electromagnetic radiation.

But now we had a problem. Waves require a medium, something to do the waving. Sound, for example, is a wave in air. No air, no sound. But we receive light from stars that are very, very distant. If empty space is a vacuum, then what is doing the waving to get the light waves from there to here? Physicists called this undetected medium the luminiferous aether and sought to detect it directly or indirectly. That is what Einstein was doing when he injured himself in the lab in college. By evacuating a glass jar, was he pulling out just air or was he sucking out the aether as well? If aether could be moved by physical forces, then the vacuum pump could remove it, causing an "aether vacuum" and making the jar no longer transparent—without aether inside the light waves could not be transmitted through it. These conditions would turn the evacuated jar into the inverse of a black hole: instead of keeping light contained within it, the jar would never allow light to enter.

Einstein grew skeptical of the existence of the aether. But as long as light had to be a wave, the aether was necessary. If physics was to be rid of this unnecessary scaffolding, then a new picture of light would be needed. Hints of such a depiction were emerging. One came to be known as "blackbody radiation."

Objects give off light when heated—think of an iron bar in the blacksmith's furnace. Now, suppose we heat up an empty metal sphere. The light emitted from the interior surface would be trapped inside, some of it reabsorbed by the walls and some reflected inside. The question physicists asked was, "If we poked a small hole and look inside, what would we see?" In other words, how much would be reabsorbed, how much would be bouncing around, and what wavelengths would we observe?

The standard understanding at the time gave the problematic result that as we look at smaller and smaller wavelengths of light, the amount of energy for those colors would get larger and larger, so that in total an infinite amount of energy would flow out of the hole. But this is impossible: you cannot put in a finite amount of energy and get out an infinite amount. From experimental data we knew what actually happened, and it was quite different. This was termed "the ultraviolet catastrophe."

We had a flawed theory and data we could not explain. A lot of effort from the smartest physicists at the time went into trying to solve this problem, but progress was slow at best. We had different models that were correct for different ranges, but no underlying account that held across the spectrum. This meant that no general sense of the underlying mechanism could be developed.

Planck decided to work backward. Instead of finding a model that fit the data, he started with the data and figured out what function would give him the observed energy curve. He found that we could account for blackbody radiation if we think of light as emitted and absorbed not as waves but as individual packets, as particles. For Planck, this was quite disturbing. Convinced that the atomic hypothesis was wrong, that matter is continuous and light is a smooth wave, he held that nature could not come in bits. But here was his finding, and so he declared that we needed to think of light *as if* it was absorbed and emitted in discrete clumps. While it was traveling, the wave description worked perfectly, but to avoid the ultraviolet catastrophe we would treat light as if it were made up of particles when it is given off or taken in. Planck took this not literally but only as a heuristic device, a mental image to help us think about light in certain circumstances.

Einstein's intuitions were the opposite of Planck's. Where Planck was disturbed by anything that was not smooth, Einstein preferred a chunky universe. He had already instantiated this picture with his work on atoms, and now he had his sights set on light. Planck was correct, Einstein thought, more accurate than he would allow himself to be. Planck gave us more than just a way of thinking; he showed us the light.

This idea of turning light into a collection of bits was not an isolated finding. J. J. Thomson and Philipp Lenard had shown that cathode rays were not rays at all, but actually collections of little particles we came to know as electrons. Lenard had also shown that light of the right color could kick electrons out of metal in a fashion that waves could not. Einstein had been greatly impressed with the work of Lenard, writing excitedly to Mileva about it before they were married.[10] If the cathode ray was actually a stream of particles, why not light?

The title of Einstein's third 1905 paper, "On a Heuristic Point of View Concerning the Production and Transformation of Light," is, to some degree, disingenuous. Unlike Planck, who explicitly stated that he was treating light as a particle during emission and absorption as just a heuristic device, Einstein thought he was providing more, that he was in fact showing something "quite revolutionary"—that light is, in fact, a particle.

The argument is classic Einstein. Like his first two papers, in which we gain insight into a phenomenon by considering it in two very different, seemingly inconsistent ways, here too we see Einstein taking seriously the idea that light behaves as a particle without completely surrendering the idea that it also behaves as a wave. Indeed, even after this paper Einstein on multiple occasions will refer to light waves and come up with thought experiments in which he treats light as a wave. This might seem problematic because particles and waves are completely different kinds of things: a particle is a thing unto itself that has a particular spatial location; a wave is a disturbance in a medium, it is a ripple that requires something to be rippled and is spread out in space. For Einstein, the two were complementary; the particulate nature was the underlying picture, but when you put a bunch of them together traveling through space, they move in a wavelike fashion.

Einstein makes his case by considering entropy again. He considers a collection of particles that are taken from a smaller volume to a larger one and calculates the increase in entropy. He then thinks about light that starts in a smaller area and increases in exactly the same way. As it turns out, the entropy equations for light and for particles have the same exact form. A coincidence—or is it? If we think of light as being made up of energetic bits, it makes perfect sense.

But "making perfect sense" gets you only as far as "heuristic device": it gets you "as if." He then asks what happens when we apply this idea to three cases that the wave theory of light cannot handle. Lo and behold, all three are explained perfectly. The most famous of these is the phenomenon discovered by Lenard that excited Einstein years earlier, the photoelectric effect.

If you take ultraviolet light and shine it on a surface, electrons are kicked out. That is not strange; in fact, with the wave theory of light it is to be expected. Metals have very loosely bound electrons in their outermost shells, electrons that are easily stripped off. That is why metals make such good wires for electrical circuits. If light was a wave, then when it hits the surface of the metal, the metal would vibrate—think of the tuning fork experiment from elementary school in which the struck tuning fork causes the unstruck one to vibrate without touching. These vibrations could liberate the loosely held electrons. No problem.

Except that when you make the light brighter you should be making the wave larger, and the larger wave will give the emitted electrons more energy, meaning they will move faster. It turns out that this is not what we see. More electrons are emitted, but not faster ones. If light is a wave, this should not happen.

But if light is a particle, this phenomenon makes perfect sense. In making the light brighter you are shooting more bits of light at the metal, all traveling at the same speed. More collisions of similar particles will mean more ejected electrons traveling at the same speed. Nice and neat, another anomaly disappears if light is a particle.

So, we have Planck's blackbody case, the entropy analogy, and the three cases including the photoelectric effect. While it seemed clear to Einstein that we needed to adopt the chunky picture of light, his results were so revolutionary that many refused to accept them until other notable scientists including the Americans Robert Millikan and Arthur Compton came up with further experimental support. Eventually it simply became undeniable in the face of mounting evidence, no matter how hard the establishment tried to deny it.

In March, April, and May of 1905, Einstein established the existence of atoms, forever changing our picture of matter, and he overthrew the universally held wave theory of light, forever changing our view of optical phenomena. Not bad for someone without a job as a working scientist. What was left to be done? Einstein had his sights set on the biggest of big game in physics: mechanics. Isaac Newton's theory of motion reigned supreme for three hundred years and was widely considered the greatest scientific advance in human history. The lowly patent clerk had his work cut out for him in no uncertain terms.

The paper that introduced Einstein's most famous work was titled "On the Electrodynamics of Moving Bodies." He begins by considering a coil of wire connected to a circuit. Now, take a magnet and move it back and forth inside of the coil. The result is a current in the circuit. Next, hold the magnet still and move the coil back and forth around it at the same rate. The result? The same current. It doesn't matter which is moving and which is still. All that matters is the relative state of motion between them.

The problem was that the best existing theory at the time, Maxwell's electrodynamics, gave completely different explanations for the two cases. In the case of the moving magnet, you have a changing, or "dynamic," magnetic field. In the case of the moving coil, you have a fixed, or "static," magnetic field, and these are entirely different cases as far as Maxwell's account is concerned.

At least, they were entirely different cases as long as you had a luminiferous aether—an underlying structure filling all of space—because then you could say which one was really moving, and the moving magnet was different from the moving coil because they were moving or stationary with respect to aether. But in the paper on light, Einstein had turned light into particles, and, unlike waves, particles do not need a medium. If light is not a wave, then we don't need the luminiferous aether, and without it we can say that the moving magnet and the moving coil are just different descriptions of the same physical situation reported from different perspectives. Once again we have the classic Einstein move of taking two different ways of looking at something and bringing them together to give us a new insight.

But there was one more problem that sprang from Maxwell's theory. When you derive the equation for light, the one that has the form of a wave equation, and you look for the term that represents the speed of the wave, it turns out to be a constant. And when the value of that constant is calculated from the theory, it turns out to be exactly the observed speed of light.

But the speed of light could not possibly be a constant. Consider parting friends, one leaving on a train as the other watches from the platform. They wave to each other as the train pulls out at two miles an hour. The person on the platform sees his friend on the train moving at two miles an hour, while the person on the train sees her friend moving the other direction at two miles an hour. Now consider someone who just got off the train and is running at five miles an hour past the person on the platform in the opposite direction of the train to greet his waiting family. The person on the platform would see this person moving at five miles an hour, but the person on the train would see the person moving at seven. Velocities add in a simple fashion.

Suppose a brother and sister are playing hide-and-seek in the dark, the seeking brother carrying a flashlight. When the hiding sister is found, the brother initially remains still but then starts to walk toward her at two miles an hour, keeping the flashlight shining in her eyes the whole time. When he was still, the light shining in her eyes was coming at her at the speed of light. When he started to approach, surely she saw the light at the speed of light plus two miles an hour. That is what both common sense and Newton tell us.

In Maxwell's theory, the speed of light is a constant, but a constant for whom? In what frame of reference is this speed constant? If two people are moving relative to each other, the speed surely changes for each. Who has the perspective of perfect rest such that the speed of light is a constant for him? As long as we had a luminiferous aether, there was an answer: the speed of light was constant with respect to the aether. Find the aether frame and you find where the speed of light is constant.

So we looked for it. The most famous experiment came from the American physicists A. A. Michelson and Edward Morley, who used light and mirrors to try to measure the Earth's motion relative to the aether. Their experiment detected nothing, which seemed to indicate that the Earth drags its aether along with it. But in an experiment that French physicist Hippolyte Fizeau designed to show that aether gets dragged, he too found no effect, seeming to show that the aether was not dragged but stationary. Either it moves or it doesn't—it cannot do both. What to do?

This question worried the Dutch physicist H. A. Lorentz, especially the Michelson-Morley result, and he worked hard to figure out how to make sense of it. Nothing worked because we seemed to have an irreconcilable conflict between Newton's mechanics and Maxwell's electromagnetism. Every attempt to fix Maxwell in a fashion that would bring his theory in line with Newton's failed. Then Lorentz found something: if we go the other way and fix Newton's mechanics to work with Maxwell, then we can account for the Michelson-Morley experiment on the condition that we treat lengths as if they shrink in the direction of motion. Of course, they don't—that would be too weird. Again, it is a heuristic device not to be taken literally, or so Lorentz thought.

Einstein had been discussing this question all day with his friend Michele Besso, after which he caught the train home to Bern. Getting off the train and walking toward his flat, he looked back at the clock on the station tower to see what time it was. Then it struck him. He was seeing not what time it *is*, but what time it *was*. To read a clock, light must bounce off the face and hands and travel to the observer. That takes time. In observing the time what you are seeing is what time it was when the light left the clock. But if the observer was moving away from the clock, that light would have to not only get to the observer who continued looking at the clock, but catch up as he walked away. As a result, it would seem that the clock was running slow.

What if Lorentz was more correct than he realized and it was not just as if the lengths contracted, but it was the case that they really did? And what if the clock actually is running slower for the person moving as a result? What would such a world look like for such observers? This would be the project he set out for himself.

Einstein started from two axioms, basic assumptions from which everything else would follow. The first is the constancy of the speed of light. Having done away with the need for a luminiferous aether, Einstein contended that we must take as a starting point that the speed of light in a vacuum is always the same for all observers, no matter their state of motion with regard to the source. If we go back to our siblings playing with the flashlight, the sister who has the light in her eyes will see that light coming at her at the same speed whether her brother is still or moving toward or away from her with the flashlight. This seems intuitively wrong, but it is the first claim that Einstein takes as necessarily true.

The second is the principle of relativity,[11] a notion he found in the writing of Henri Poincaré, according to which the laws of physics should be the same for all observers who are moving at a constant speed in a straight line with respect to each other. This constraint comes from the fact that when people accelerate, forces appear in one frame but not in another. Think about riding a roller coaster. Going up, you feel a pull back in your seat. Going down, you feel as if you are being pulled up out of your seat. Turning a corner throws you from side to side. To you, there are forces experienced. Your friend watching from the ground would chalk it up to your momentum going in a straight line and the car accelerating around you as it follows the track. What is a force in an accelerating frame is not necessarily seen as a force from another. This difference in forces requires an adjustment in how the physics accounts for them. But if we consider only frames of reference moving in straight lines at constant speeds relative to one another, then there will be none of these "factitious forces," as Einstein termed them, so the physical description for different observers ought to be the same.

Einstein is careful in setting out this principle to specify that the laws of physics he is talking about are the laws of electrodynamics and optics—this means Maxwell's laws with their specification of the constancy of the speed of light. Newton and Maxwell cannot both be right, and he is making clear that he is buying Maxwell's story and not Newton's. This is what undermines Newton's mechanics and thereby requires a new theory of motion.

That theory of motion is precisely what comes out when we combine these two postulates. Einstein derives what we now call the "Lorentz transformations," which determine how lengths and durations will be measured by different observers and show the relationship between them. Just as Lorentz argued, lengths contract when measured by observers who are moving with respect to the thing being measured. The closer the thing gets to the speed of light when measured from your reference frame, the thinner you would measure something as being. If you were traveling the speed of light, the lengths of all things would squish to nothing, but only in the direction you are traveling. Things would still have the same height and width, but their lengths would disappear.

You are moving relative to other observers, but you remain still according to those things you carry along. No matter how fast you move, the things you hold—because they are moving at the same speed in the same direction—will not squish according to your measurements. But when the other observers measure those things, they will see them as having shrunk. The question of length becomes purely one of perspective. There is a fact as to the length of an object in a reference frame, and Einstein's and Lorentz's equations show how that length changes when measured in different reference frames. Length ceases to be something absolute and universal; it becomes a matter of your relative state of motion when you are measuring it.

Similarly, the faster you travel the slower time passes compared with an observer at rest. If you had a watch and I had a watch, and you were traveling close to the speed of light, then my watch would seem to run slow to you. If you were traveling at the speed of light, then time in my reference frame would be seen by you to have stopped completely. I would never age in your eyes. Similarly, your watch would seem to stop when I view it. As with length, this measure is related to reference frame.

This alteration in our measurements of spatial distances and temporal durations plays havoc with our understanding of motion. Velocity, after all, is the distance something moves in a unit of time, so changing measurements of distance and time will change how speeds are measured from one reference frame to another. Think back to our parting friends at the train station. We have one person on the train pulling out at two miles an hour and her friend watching from the platform, while someone who just got off the train walks in the opposite direction from the train at five miles an hour. The woman on the train would see her friend fading away from her at two miles an hour while the person walking past the friend would seem to be moving at seven miles an hour. According to Newton, velocities simply add. Similarly, when we considered the case of our kids with flashlights, if velocities add as Newton demands, then when the brother is walking with the flashlight toward his sister, the speed of the light she observes should increase by the speed of his walking.

But by the first postulate this cannot be true according to Einstein. We need a new way to add velocities to make sure she sees the light moving at the same speed no matter how fast he moves with flashlight toward her. Einstein derives this and shows that it too depends on the speed of the observer relative to the speed of light. The woman on the train will see the person walking past her friend moving at slightly less than seven miles per hour, yet the difference is so small at those speeds that the change is not detectable. But the closer to the speed of light you go, the more that difference matters. At the speed of light, the contribution from the motion disappears so that any velocity plus the speed of light turns out to be the speed of light again.

In the second part of his paper Einstein says to his readers, OK, I've given you a strange new way to see the physical world, but now here's how to test if it works. Just as in the light paper, Einstein sets out several different physical effects that could be explained only by the new theory. He considers the optical Doppler shift and the effects of reflected light, but it is in thinking about moving electrons that something new and strange is predicted by the theory. Not only do we find that observers in different states of motion measure a difference in length, time, and velocity—things we thought with Newton were absolute facts of the world—but they would also measure differences in the mass of the electron.

The faster it goes, the heavier it gets. At the speed of light, the electron would be infinitely heavy. This would give it an infinite amount of kinetic energy. This cannot be, of course, and so Einstein is led to assert that "velocities greater than that of light have—as in our previous results—no possibility of existence."[12] The speed of light is not only a constant, it is also a limiting velocity. Nothing can move faster than this speed. It is not an engineering problem; it is not that we have yet to figure out how to do it. If Einstein's theory is correct, then moving faster than the speed of light would require an infinite amount of energy, and that is not possible. Nothing can move faster than light in a vacuum.

This last result about mass led to Einstein's final revolutionary paper of that miracle year, "Does the Inertia of a Body Depend upon Its Energy Content?," in which he sets out no new theory or conceptual framework. It is just a short note, yet it contains his most famous result. In a few paragraphs he summarizes the first relativity paper and then goes on to point out what would have to be the case with respect to a body that emitted light when viewed from different reference frames. It turns out, for reasons related to the change in mass of the moving electron, that the energy of the body that gives off the light is reduced by an amount that has nothing to do with the composition of the body.

Einstein writes that "*if a body gives off the energy L in the form of radiation, its mass diminishes by* L/c^2*.*"[13] But this wrongly seems to be related to the fact that the emitted energy is light. "The fact that the energy withdrawn from the body becomes energy of radiation evidently makes no difference, so that we are led to the more general conclusion that 'The mass of a body is a measure of its energy content.'"[14] Since the energy need not be in the form of light, we can change the L to its usual E and assert that $m = E/c^2$, or doing some basic algebra allows us to put it in its iconic form, $E = mc^2$.

The mass of the body, of anything that has mass, is a measure of its energy content. This means that mass is a form of energy. On one hand, we know that energy comes in many forms—such as light, heat, motion—and we have long known that we can change one form into another. But energy was seen as a particular physical quantity quite different from mass. The two notions are in different categories in the same way that something has both size and color, but to say that something is blue is to say nothing about how big it is. Here is Einstein arguing that as part of the fundamental structure of the universe, the two are the same sort of property.

The world as we thought we knew it was constructed by Isaac Newton from basic concepts: space, time, motion, mass, and energy. In his masterwork *The Mathematical Principles of Natural Philosophy*, Newton begins by defining these notions or, in the case of space, time, place, and motion, giving a brief discussion of them since they need no defining, "being well-known to all."[15] From these basic notions, Newton developed three elegant laws of motion and a law of universal gravitation that explained the falling of apples and the orbits of planets, the motion of comets and the rising of the tides. It was so successful that it was thought the highest expression of the human mind in all recorded history.

Yet here was a mere patent clerk, a civil servant who could not secure even the lowest ranking research position, claiming to have a new set of concepts that must replace those that had served us so well for hundreds of years without fail. In one year, this scientific nobody had contended that observable facts will force us (1) to radically change how we understand the nature of matter, establishing the atomic hypothesis around since classical Greek times, and establishing a controversial picture of heat, (2) completely change our understanding of the nature of light, thus eliminating the luminiferous aether that seemed essential both to the standard understanding of Maxwell's theory of electricity and to magnetism, and (3) reject the Newtonian concepts of space, time, motion, and mass, all of which sit at the foundation of the most successful theory in scientific history, and replace them with counterintuitive notions that give rise to weird, unobserved effects. In other words, there was virtually no single part of the study of physics, the oldest and most established science, which Einstein did not seek to completely overhaul in 1905. As far as Einstein was concerned, after his work of that miracle year, *everything* was different.

3

The Happiest Thought

Albert Einstein never lacked confidence. He was certain that his work of 1905 was not only correct but revolutionary. The appropriate response on the part of the academic scientific community to such ideas would be to engage them vigorously and rigorously. He would divide the world of physics into those following him in the overthrow of the old views and those steadfastly defending them. He would become a hero to some and a serious threat to others. In his sister's words, "The young scholar imagined that his publication in the renowned and much-read journal would draw immediate attention, he expected sharp opposition and the severest criticism."[1]

Among those who championed his view, the more established and better connected would shepherd him into the academy. In Einstein's mind, he surely deserved an academic post. He would leave the patent office and become a professor at a university, where he could be a respected part of the world of professional physics. He had reshaped our entire view of material reality and he would be celebrated for it across the entire scientific community.

Indeed, this is exactly what happened—but not in 1905. He expected the reaction to his work to be significant and immediate. His biographer Abraham Pais wrote, "He was very disappointed. According to his sister, his publication was followed by an icy silence."[2] There were no letters or articles attacking him, no references to his work in the papers of other physicists, no offers of university jobs. The only thing that came in short order from his work was a 1,000-franc raise accompanying his promotion to patent examiner expert class II for having received his doctorate. Einstein's miracle year did not miraculously change his life. Revolutions can take time.

Yet Einstein's work did not go completely unnoticed. Gradually he began receiving letters and, later, visits from well-regarded, even important physicists. These connections gave Einstein hope and a sense that he was joining the club, especially when he received and was asked for offprints.

In the days before photocopiers, printed material was expensive, hard to come by, and therefore extremely desirable. If you were fortunate to have access to a library with the journals of interest in your field in its holdings, you could read the latest articles there. Otherwise you were out of luck unless you could afford to pay for personal subscriptions. Access to information that we enjoy now was generally out of the question. But when a journal printed its run, it made extra copies of each article and sent them to the authors, who could distribute them as they wished, allowing recipients to have their own personal version of the work to read and reread. Authors had a limited supply of these offprints of their works, making them extremely valuable. They became the baseball cards of the academic set. To receive one or a request for one from someone well known was a great honor. Einstein began sending some of his to the more prominent scientists with whom he corresponded and receiving some in return.

These included Einstein's heroes and influences, such as Philipp Lenard, the Nobel Prize winner in 1905 whose work on the photoelectric effect so impressed Einstein and spurred his own work on light. In correspondence with a friend, he would refer to Lenard as "a great master and an original mind."[3] He would correspond with H. A. Lorentz, whose length contraction hypothesis was at the heart of the theory of relativity in 1905. Also, his work found an admirer in Max Planck, whose introduction of the notion of the quantum filled out Einstein's views on light. Planck was the greatest supporter Einstein could have imagined. He was universally respected as the leading voice of the physics community at the time, and when he published a short article in 1906 extending Einstein's work on relativity, the topic became legitimate in the eyes of mainstream working researchers. Right or wrong, if Planck thinks it warrants being taken seriously, then it must be something worth discussing.

But as one would expect, this discussion had voices of quite different tenors. Some were supportive, others adversarial, and some confused, and yet others expressed a wait-and-see attitude. One famous exchange took place between Lorentz and Arnold Sommerfeld, an important theoretician from Munich, who wrote, "But now we are all longing for you to comment on that whole complex of Einstein's treatises. Works of genius though they are, this unconstructable and unvisualizable dogmatism seems to me to contain something almost unhealthy. An Englishman would scarcely have produced this theory: perhaps it reflects, similarly as with Cohn, the abstract-conceptual character of the Semite. I hope you will succeed in imbuing this inspired conceptual skeleton with real physical life."[4] While this passage contains some of the seeds of later antisemitic attacks on Einstein's work, Sommerfeld would become a friend and coauthor with Einstein, championing the new crop of relativity-supporting young physicists in their search for jobs in the established departments of the German universities. Ironically, he did so to such a degree that Sommerfeld himself would be slurred by Nazi-sympathizing Aryan physicists as Einstein's "business manager."

Planck's assistant Max von Laue also found Einstein's work fascinating and engaged in spirited discussion about it and its interpretation in these early days. During the slow summer season, Laue arranged his travels to spend some time meeting Einstein. Leaving the University of Berlin, one of the most prestigious institutions in Europe, he found his way to Switzerland. Since their letters were purely about matters of scientific interest, Laue arrived in Bern assuming he would be meeting another professor who was in the physics department of the university there. He was shocked to find that he was actually meeting a civil servant, a patent clerk. In fact, when Laue stepped off the train he let Einstein walk past him, certain that the young man on the platform could not be the eminent scientific mind who created the theories he came to discuss. But he was indeed, and after the misunderstanding was cleared up the two walked and talked excitedly, forming a friendship that would last for decades.

One of the most important early responses to the theory of relativity came from one of Einstein's old college professors, Hermann Minkowski. A shy man who would nervously stammer in the classroom, Minkowski was listed by Einstein in his "Autobiographical Notes" as one of his great professors[5] despite the fact that Einstein spent significantly less time than he should have in Minkowski's classes. Minkowski famously referred to his student as a "lazy dog" and expressed incredulity that it was his former student who developed this theory.

Minkowski was a Russian Jew whose family relocated to the city of Königsberg while he was a child. Here he met another young man who shared his passion for mathematics,

David Hilbert, and the two became inseparable despite their contrasting personalities—Hilbert was funny and outgoing. They would meet regularly under an apple tree with a young math professor named Adolf Hurwitz for long walks during which they would discuss questions across the entire range of mathematical fields. Hurwitz had been a student of the great Felix Klein, a towering figure in nineteenth-century mathematics who completely reshaped the modern understanding of geometry, and as a result Minkowski and Hilbert received a deep understanding of the mathematical treatment of space.[6]

Hilbert would become a rising star, receiving promotion after promotion, and he was always sure to use his newfound power in mathematical circles to pull Minkowski into each position he vacated. This ended when Minkowski secured a position at the ETH, where he would occasionally see Einstein in his classroom. But Hilbert kept on climbing, landing ultimately at the university in Göttingen, which Klein had built into the strongest mathematics department in the world. Hilbert was able to lure Minkowski away from Switzerland to join him, and the two would spend Minkowski's last years together exploring questions of mathematical physics, especially relativity.

Minkowski requested an offprint from Einstein, as he and Hilbert were offering a jointly taught seminar on relativity. Minkowski eventually published a couple of papers on the topic and gave a famous talk called "Space and Time." Minkowski was astonished by the content of the theory but thought that Einstein's lack of mathematical seriousness hid the theory's true meaning. Using the mathematical tools that Klein developed and Hurwitz had taught him, Minkowski gave a geometrical interpretation of the theory, framing it in terms of a four-dimensional space-time manifold. To truly understand the nature of reality itself, he contended, we had to free ourselves from our picture of a three-dimensional space separate from a flowing one-dimensional time and accept that we live in a four-dimensional reality. "The views of space and time which I wish to lay before you have sprang from the soil of experimental physics, and therein lies their strength. They are radical. Henceforth space by itself, and time by itself, are doomed to fade away into mere shadows, and only a kind of union of the two will preserve an independent reality."[7]

Einstein read Minkowski's work and found himself annoyed by it. Einstein worked from physical intuitions. The mathematical formulation was important, but he thought Minkowski was eliminating the pictures for abstruse mathematical structures. Minkowski may be creating a mathematically elegant version of the theory, but he was not disclosing the underlying reality; to the contrary, he was obscuring it. At one point he called Minkowski's work "superfluous learnedness,"[8] and complained that "the people in Göttingen sometimes strike me not as if they wanted to help formulate something clearly, but as if they wanted only to show us physicists how much brighter they are than we."[9] In one sense, this was not false. In disparaging Einstein's mathematical abilities Hilbert famously quipped that "every boy in the streets of Göttingen knows more about four-dimensional geometry than Einstein,"[10] but Einstein came to realize that Minkowski's four-dimensional formulation did expose the real properties of the universe itself. Minkowski did not eliminate the pictures but provided a new one that would sit at the heart of his extension of the ideas over the next decade.

Relativity did more than just spur informal discussion and a few scattered published articles; it also became a part of the work of a handful of major experimentalists. The last section of Einstein's first relativity paper focused on the application of the theory to moving electrons. His was not the only theory about this question, and the great experimenter Walter Kaufmann had been taking careful measurements of fast-moving electrons in magnetic fields to settle the question. This was data that could make or break Einstein's view.

Kaufmann's work was widely anticipated and publicly presented. He compared his results with the three central theories of the time and found that, of the three competitors, Einstein's predictions were the furthest from what he actually observed. In his words, the theory of relativity had been "liquidated."[11] Einstein, confident as always, thought the interpretation overstated and the data equivocal, and that further investigation of the matter would vindicate his view. He had little grounds for such a belief, but, it turns out, he was right.

One place he thought supporting evidence could be found was in the research of Johannes Stark, who was studying so-called canal rays, which were atoms that underwent significant acceleration in an electrical field. Atoms have "fingerprints," meaning telltale wavelengths of light they emit very precisely. Stark found that in an electrical field those lines shifted uniformly in color. This is a version of the Doppler effect we all experience when an ambulance speeds by: in the same way that the wavelength of the siren changes so that we hear the pitch get higher as the vehicle approaches and lower as it zooms away, so too the light coming out of the excited atoms undergoes the same sort of shift. Einstein thought that a careful measurement of these from the fast-moving atoms could give evidence of the time dilation effect predicted in the first of the 1905 papers on relativity. Stark was a talented and meticulous experimentalist who won the Nobel Prize for his work in 1919, but at the time his measurements were not precise enough for Einstein's needs.

Not only was Einstein an admirer of Stark's, but Stark also was impressed with Einstein's ideas, which he publicly defended against Kaufmann's conclusions. Stark had founded a journal, the *Yearbook of Radioactivity and Electronics*, and he invited Einstein to write a review article on relativity for an upcoming volume in 1907. Academic journals print three types of articles—(1) positive articles, which make a claim and give supporting evidence,

(2) negative articles, which give counterevidence against someone else's claim, and (3) review articles, which give the lay of the land in a particular subfield: explaining the basic notions, discussing the work of the important active researchers, and setting out the various points of controversy and open questions being pursued. Stark saw relativity as a burgeoning new area and thought that Einstein's account of its foundations and where it was going would be valuable. Einstein reluctantly agreed, his hesitation coming not from his lack of confidence in the theory or his excitement about the others working in the area, but because as a patent clerk in Bern he did not have access to the library resources needed to do a full literature search for all of the research interest that was under way, especially given the short two-month deadline he was given.[12]

Writing the article, Einstein found himself dissatisfied with the theory in two ways. First, he realized that his 1905 papers touched every part of physics—except one. His work on atoms got to the basic constituents of matter and accounted for the concepts of heat in thermodynamics. His work on light revolutionized optics, and relativity's use of Maxwell's work allowed electricity and magnetism to be represented as well. Finally, he had used these notions to reformulate mechanics, the theory of motion. But one force was missing, as completely absent from Einstein's work was gravity. Why should this one force not show up in this new understanding of the universe?

The second problem was the limitation inherent in the principle of relativity. Einstein's work connected the reference frames that were in motion relative to each other at a constant speed in a straight line. This limitation seemed arbitrary. The true laws of physics should not care if you are walking straight ahead or waltzing on a rotating disk. The laws are the laws, and a complete accounting for the underlying principles governing the behavior of the universe should be "generally covariant"—that is, they should hold for every single observer no matter his or her state of motion. Einstein would need to generalize his theory of relativity.

But how to do this? Einstein had two problems, gravitation and acceleration. How could he solve them? Working at the patent office, he had one day what he called his "happiest thought." It was a simple realization. If a person fell off a ladder, during his fall he would not feel his weight. When a person is in free fall, it is the same as being weightless. For example, if you are standing on a scale in an elevator in which the cable has snapped, the scale will read your weight as zero because you and the scale are accelerating downward at the same rate. Similarly, stand on a scale in a rocket far out in space, away from any massive object, and again you will be weightless. Acceleration in a gravitational field and being stationary where there is no gravitational field result in exactly the same observable effects.

Just like the magnet and coil in his relativity paper, the fact that the results are the same must mean that they are just different descriptions of the same physical phenomenon. Here we see yet another instance of Einstein taking what seem to be two separate ideas and considering them to be just different ways of seeing the same thing from different viewpoints in order to find new and exciting insights by combining the perspectives. Gravity and acceleration were not two problems to be solved, they were just different elements of the same problem. He would need a generalized theory of relativity that accounted for the "factitious" forces that are experienced in accelerated reference frames—for instance, what you feel in your car when you step on the accelerator or the brake, or suddenly turn the steering wheel—and the pull of gravity as different instantiations of a single physical effect.

But what kind of effect? Where did it come from? The clue came from Minkowski's work. Minkowski had shown Einstein that he needed to think of the theory of relativity as setting out a structure of a four-dimensional universe in which space and time were combined into a single entity with a definite internal geometric structure. Could we think of gravitation as affecting that structure in such a way that would seem similar to the ways in which that structure would be seen differently by someone speeding up and rotating through it? Maybe, but figuring out the mathematical description of something like that would be hard, really hard. It would, Einstein realized, make the theory of relativity he had developed in 1905 into mere child's play. Einstein had a new project, one that would take him a while.

Between the correspondences, visits, and other professional attention he and his work were receiving, and because he'd been publishing numerous reviews and other articles, Einstein felt himself becoming more and more a part of the scientific community, even if he was in its outer orbit. To really be in the thick of it, he needed a university post. As with so much else, Einstein assumed the rules did not apply to him. Rather than the usual process, a professorship would simply appear for him because of the strength of his ideas.

The universities of the time were government institutions at which the appointment of professors was a matter of a central bureaucracy, and, as is the case with bureaucracies, there was a procedure. After you received a degree and a doctorate, you became a *privatdozent* at a university. This is where you paid your dues. As a privatdozent, you were considered an adjunct member of the department but received no pay. You were expected to give regularly scheduled classes that augmented the lectures of the professors, but the students would pay the instructor directly—that is the "privat" part. If you were a good teacher lecturing on relevant subjects, you would have lots of students and would get both money and a reputation. In addition to teaching, there was a research component and you were expected to write a *Habilitationsschrift*, a written work that showed you to be a cutting-edge researcher. In modern terms, the doctoral dissertation was what we now think of as a master's thesis, while the Habilitation was what we now consider a dissertation.

Einstein hoped to bypass this step but was unsuccessful. Professor Kleiner (through whom Einstein received his doctorate at the University of Zurich) thought he could get a newly created position in theoretical physics in Zurich. Physics at this time was an almost exclusively experimental field, but the sort of work that Planck, Einstein, and others were doing seemed to be a wave of the future and Kleiner sought to have a dedicated position for it at Zurich. When he approached the bureaucrats, the best he could get was not a full professorship, but a new assistantship underneath him. He thought that Einstein would be a good fit for that spot, but to get that appointment Einstein would have to play the game and first become a privatdozent. Kleiner convinced Einstein that it would be prudent to get himself such a post at the university at Bern.

Einstein did, and he held his first lecture series on the physics of heat. Since he needed to maintain his job at the patent office, he held his lectures on Saturday mornings at 7:30. They were attended by four students—three of Einstein's friends and one university science student. After a while his friends faded away, leaving him with just a single attendee for his initial class.

Kleiner told Einstein that in order to make the request to bring him to Zurich as an assistant, he would need to observe Einstein so he could report on his teaching. Traveling to Bern, Kleiner observed an Einstein lecture that was a complete disaster. Convinced that Einstein would be a menace in the classroom, Kleiner wrote a report that not only cost Einstein the appointment, but began a widespread whisper among Swiss and German physicists that Einstein was unsuited to teach at all.

Enraged by his growing reputation as pedagogically incompetent, Einstein contacted Kleiner and demanded that he get invited to Zurich to lecture to the students there to show that he was not the disorganized, rambling, incomprehensible monologist Kleiner thought him to be. Kleiner agreed, and this time Einstein gave an unusually good presentation, good enough to convince Kleiner to support him.

The process of getting Einstein an official offer was not completely smooth. Kleiner had to convince those at the university, and this meant assuaging the antisemitic concerns of his colleagues. Kleiner made clear to them that he knew Einstein well enough to attest that while Einstein was of a Jewish background, he did not display the usually associated disagreeable characteristics they believed Jews to possess.[13] His word was sufficient on the matter, and the offer was made.

But an offer is not a job. The proposed salary was less than he was making as a patent clerk. Financial matters were of real concern to Einstein because he had a family to support, so he turned down his chance to become a physics professor. It is not clear whether this was simply a negotiating tactic or if Einstein truly was willing to walk away from the opportunity to become a working scientist and remain a bureaucrat for life over a few francs. But the offer was ultimately increased to equal his salary at the patent office, and he accepted. Albert Einstein was now a professor.

The move to Zurich in 1909 was good for the family. Einstein and Mileva were back in the city they loved and where they had an established network of old friends with whom to talk, eat, and play music. Soon there was another addition to the family, as Mileva gave birth to their second son, Eduard, in 1910. The labor was difficult, and Mileva's recovery was slow. She brought her mother into the household to help with the children and housework. Einstein met his students at a local coffeehouse trying to re-create the magic of his own university days. He built intricate toys out of matchboxes and string for the kids. Life was happy, but it was not to last.

Einstein received an inquiry from the German University in Prague about his interest in a full professorship there. A chair had been created for theoretical physics, and Einstein would be not only a good fit but also an impressive addition to the faculty. For Einstein, it would be a major step forward in his career. At Zurich, he was a mere assistant; at Prague he would hold his own position. Einstein responded, clearly indicating his interest.

The offer was not a straightforward matter, however, for two reasons. First, the application required the candidate to specify a religious affiliation, and being a nonbeliever automatically disqualified one from employment by Emperor Franz Joseph's government. Einstein was reluctant, having renounced his connection to Judaism. Eventually he agreed to have "Mosaic" put down on the form. Secondly, he would have to become a citizen of the Austro-Hungarian Empire, but he refused to give up his Swiss identity. Once again, nationalism and religion would complicate Einstein's life. A compromise was reached, and Einstein was allowed to maintain dual status. The matters resolved, Einstein joined the university in late 1910.

The university gave him a nicely appointed office, its major quirk being that it looked down into the yard of the adjacent insane asylum. Of its inmates Einstein would say, "There are the other crazy people, the ones who do not work on physics."[14]

Enhanced professional and social status and a significant raise came with the new position, but the move had unexpected negative consequences. The formality of the university conflicted with Einstein's demeanor. His disdain for authority led him to dismiss the usual class-based etiquette that was expected. Einstein biographer Philip Frank noted that "the tone with which he talked to the leading officials of the university was the same as that with which he spoke to his grocer or to the scrubwoman in the laboratory."[15] Indeed, he got off on the wrong foot in this way. When informed that it was standard practice for incoming faculty members to pay personal visits to the homes of the other professors, Einstein at first was willing to do so. He visited those who were closest or who lived in a section of town in which he wanted to stroll. But once the task began to become tedious, he simply stopped, thereby snubbing significant colleagues who thought they deserved such a visit.

Einstein's combined loathing for mindless social formality, coupled with his sense that his importance put him above it, led him to adopt a strange combination of arrogance and empathy. Einstein could be silly, poking fun at the absurdity of social custom, sarcastically undermining those who held social power. Frank wrote:

> Einstein's conversation was often a combination of inoffensive jokes and penetrating ridicule, so that some people could not decide whether to laugh or to feel hurt. Often the joke was that he presented complicated relationships as they might appear to an intelligent child. Such an attitude often appeared to be an incisive criticism and sometimes even created an impression of cynicism. Thus the impression Einstein made on his environment vacillated between the two poles of childish cheerfulness and cynicism. Between these two poles lay the impression of a very entertaining and vital person whose company left one feeling richer for the experience. A second gamut of impression varied from that of a person who sympathized deeply with the fate of every stranger, to that of a person who, upon closer contact, immediately withdrew into his shell."[16]

Einstein's complicated personality put him at odds with his surroundings, alienating those who thought themselves powerful. "Persons who occupied an important social position frequently had no desire to belong to a world whose ridiculousness in comparison to the greater problems of nature was reflected in this laughter," Frank noted. "But people of lesser rank were always pleased by Einstein's personality."[17] In the lofty realm of academia, Einstein found himself in a world he viewed with scorn. The seemingly endless faculty meetings struck him as both trivial and vicious. The forms and rules of the university bureaucracy used up valuable time and energy.

Additionally, Einstein found himself intellectually cut off. There were no other theoretical physicists with whom to discuss his thoughts. The assistant he had brought with him moved on, and the replacement was not up to Einstein's standards. He felt stranded just as his theoretical musings were becoming more and more difficult.

The situation was even worse for his family. Living a purely domestic life, Mileva was becoming more and more alienated from Einstein, who saw her as a homemaker and not as a scientific colleague. With the move to Prague, Mileva not only lost her friends and social network, but she stepped into a part of the Austro-Hungarian Empire which harbored racial bias against her as a Serb. Unlike the cosmopolitan Zurich, which Mileva had never wanted to leave, the people of Prague had a different character that struck the Einsteins as snobbish and uncaring. Though they found some friends in the new city (Einstein spent some time in the circle of prominent Jewish intellectuals that included Franz Kafka),[18] Mileva wanted to leave. She was miserable. Einstein was absorbed in his work to the point of neglecting his family. Eduard was ill; he would be physically and mentally unwell for most of his life, and Mileva believed that the water and air of Prague were partly to blame. Life was getting darker and their marriage was getting rockier.

By 1911, Einstein's place in the world of theoretical physics had been well cemented, and this made it possible for him to look to relocate out of Prague. He had received an offer from the Dutch university at Utrecht, but Mileva wanted to return to Zurich. The one attractive aspect to the Utrecht position for Einstein was its association with his hero Lorentz, whom he would meet that year at the initial Solvay Conference in Brussels.

Ernest Solvay, a Belgian chemist who made a fortune from

his patented means of producing soda ash, decided to bring together the top scientific minds from across the world to discuss a cutting-edge topic of interest, radiation and the quantum. Marie Curie, Max Planck, Ernest Rutherford, and Jean Perrin were four of the ten in attendance who were or would become Nobel laureates. It brought together people Einstein knew and respected and whose work he had read in admiration for decades. His great influences, Lorentz and Poincaré, would both be there. Even if he had not felt intellectually isolated in Prague, this gathering would have seemed a dream come true. It was a who's who of the scientific world, and Solvay made sure that they were treated like royalty, something highly unusual for academics, even those at the elite level.

Little of consequence emerged from the conference itself, and Einstein thought it largely a waste of time. He did, however, find himself even more taken with Lorentz than he anticipated. "H. A. Lorentz presided with incomparable tact and incredible virtuosity. He speaks all three languages [German, French, and English] equally well and has a uniquely acute scientific mind and delicate tact. A living work of art."[19] His interactions with Poincaré, on the other hand, were more of a disappointment. He had hoped to convince both of the forefathers of relativity of his idea's veracity, but while Lorentz found himself on the fence (a few years later Einstein would be invited to give an address at Leyden, where he would again try to convince Lorentz), Poincaré was dismissive to the point of stubbornness, something that annoyed Einstein, who, of course, was more than capable of displaying the same trait when a view contradicted a deeply held intuition.

Traveling from the conference allowed Einstein to stop in Utrecht to discuss the proposed position, one of several inquiries from various universities. He was not serious about Utrecht, but thought that a competing offer might give him some leverage with the Swiss authorities in his bid to return to Zurich.

The ETH, Einstein's alma mater, had recently been given the authority to grant Ph.D.'s, and with the influx of graduate students came a simultaneous increase in the research expectations of the faculty. No longer were they just teachers, as the institution now needed world-class scientists working at the front line of the discipline. When an opening occurred for a physicist, Einstein thought himself perfect, especially since the job required no teaching of large introductory classes, only smaller seminars with advanced students. It was his old university. It was the Zurich that he loved and to which Mileva insisted on returning. Einstein wanted the job.

Weber was still in charge of the physics department, and Einstein had burned that bridge long ago. But Weber was older and in poor health and would therefore not be a part of the search for the new colleague. Einstein needed help from the inside. Fortunately, his old friend Heinrich Zangger was at the university, and the head of the entire division in which theoretical physics was housed was none other than Einstein's university chum and guardian angel, Marcel Grossmann. Grossmann had finished at the ETH with Einstein and went across town to the University of Zurich to earn his Ph.D. in mathematics. He returned to the ETH and had been teaching there for several years.

Einstein appealed to Grossmann and Zangger to pull all the strings they could, and the behind-the-scenes machinations were well under way when an inquiry about his interest in the Utrecht position arrived. Ordinarily, this would have been of little concern, and Einstein would simply have thanked them for their interest and declined politely. But this note was from Lorentz himself. Einstein wrote to his operatives that time was of the essence. If Lorentz contacted him again directly he might not be able to decline at the risk of insulting him. But if he could write back that he had already accepted another position, he could save face with his hero. Grossmann, who had smoothed the way so that Einstein could get through the ETH in their college days, and who then got him the job at the patent office when he needed it, once again came through and secured for Einstein the position back at the ETH. Einstein gingerly addressed Lorentz: "I write this letter to you with a heavy heart, as one who has done a kind of injustice to his father." He explained that when he left Zurich for Prague he had given his word to the Swiss academy that he would return when a suitable position opened and that one had. So in 1912 the Einstein family moved back to Zurich.

If Einstein thought that their return would reintroduce harmony into the household, he was wrong. Mileva's moods got darker. She was depressed and jealous of anyone or anything that attracted Einstein's attention. It did not help that Einstein not only was spending an incredible amount of time working on his general theory of relativity, but that he was also rekindling a relationship with his cousin Elsa. Mileva's state drove Einstein away, and Einstein's alienation only made things worse for Mileva. It was a domestic spiral heading downward. She also developed rheumatism, which made it difficult for her to leave the house. So even though she was back in Zurich, she could not visit her friends. Mileva was stuck in the house, stuck in her life, and her depression got worse.

Things were also bad at work. The mathematics that would be needed for the general theory of relativity included new and advanced tools Einstein had never seen. Minkowski had shown Einstein that he would have to think in terms of the geometric structure of a four-dimensional space-time. This was tough enough. But Einstein realized that if he were to impose a gravitational field into this geometry, it would require a complex set of mathematical entities, called tensors. These would be the basic component of all of the calculations and would turn the space-time of his theory away from the Euclidean geometry he fell in love with as a child and into a much more complex non-Euclidean space that he had no idea how to work with.

Einstein needed someone with the technical background to teach him what he needed, someone who could understand the physics and who already knew the mathematics, someone who could sit down for long stretches of time and work with him through the new computational tools. It just so happened that Grossmann's doctoral dissertation and subsequent research were in this exact area. When Einstein described his ideas about a relativistic gravitation theory and how it would cause space to become warped, he asked Grossmann what sort of geometry would be required. Grossmann explained that it was the geometric approach developed by Bernhardt Riemann, a new generalized picture of geometry, but that Einstein should be worried because it was perhaps a bit too difficult for physicists to understand.[20] When Einstein asked if there was a way around this, Grossmann said no, and so, just like a decade earlier when the two studied from Grossmann's notes for their university exams, they came back together at the ETH and worked from Grossmann's research to bring Einstein up to speed. The two began to work on figuring out what sort of equations would be required for Einstein's task. For two years, they collaborated closely on papers that explored—sometimes correctly, sometimes incorrectly—the structure of the new theory.

While Grossmann was the only person in Zurich with whom Einstein could discuss the technical details, there were plenty of colleagues at the ETH who were constantly dropping in on Einstein to discuss his work. This had something to do with his fame, but not in the sense one might have expected. A strict no-smoking policy had been put into effect for Einstein's building, but since Einstein was Einstein, he assumed the rules did not apply to him and word soon got around that Einstein's office was the place to go if you needed to light up. As such, several times a day every smoker in the physical sciences developed a sudden interest in the theory of relativity.

In July 1913, Einstein was honored to receive a visit from Planck, accompanied by Walther Nernst, who had spearheaded the Solvay conference and would go on to win the Nobel Prize in chemistry as a founding father of physical chemistry. The two came to Zurich not for a social call, but to recruit Einstein to Berlin. Planck was assembling the top physics community in the world around the university, the Prussian Academy of Science, and the Kaiser Wilhelm Institute, and he wanted Einstein to be a part of it. Einstein was flattered by the offer and extremely excited by the details—he would be very well paid as an associate of the university, the academy, and the institute, but would have virtually no expected duties. He would have the right, but not the expectation, to teach—no more administrative paperwork, and no more faculty meetings. Planck was offering Einstein a prestigious position with a large salary that required him to do only the thing he wanted to do most, work on his general theory of relativity. He would be surrounded by colleagues he liked and respected, and with whom he could have the sort of detailed discussions of his work in progress. It was Einstein's dream job.

But Einstein's dream was Mileva's nightmare. She refused to uproot the family again, and she refused to leave Zurich, especially for Germany. Switzerland was neutral territory, it was their joint home, the place she and Einstein had met and created a life together. Germany was his homeland. There they would be surrounded by his family, which never really embraced her. The prospect made her angrier and angrier. Einstein, incensed by Mileva's dissent, gave her an ultimatum—he was moving to Berlin, and if she wanted to remain married, she would come too. Ultimately she relented, and they moved in 1914, but while Mileva moved in body, she did not relocate in spirit. Her heart was still in Zurich, and the ramifications of the move became immediately apparent in their marriage—the arguments were constant and the resentment between them became unbearable. They separated, and Mileva and the boys returned to Zurich.

Einstein was deeply affected by the breakup of his marriage, and he desperately missed the boys. But with the move to Berlin, his relationship with his cousin Elsa deepened. Elsa was his cousin on both sides. Elsa's mother was Einstein's aunt, and Elsa's father was Einstein's father's first cousin. She loved Einstein and doted on him. His everyday needs were taken care of, as she treated him like a prince. Elsa made life extremely comfortable and freed him from the mundane tasks of life to focus on his work.

Work was moving forward, but slowly. The previous year Einstein had presented a version of his and Grossmann's work at a conference in Vienna, but the talk was not well received. It generated, in his words, "very lively discussion," but not in the supportive sense. Gustav Mie, a colleague of Johannes Stark's at the University of Greifswald, was enraged that Einstein had failed to mention his work at all. Einstein was deferential, knowing there were problems in his own formulation, but Mie told the gathered scientists that he was about to publish a paper that would show a fundamental and fatal flaw in Einstein's relativity theory.[21]

Einstein did not think that Mie's paper would be a threat. Yes, he was having problems with the gravitational version of relativity, but he thought Mie's objections would not be decisive. He was correct; not only did Mie's work fail to refute Einstein's general theory of relativity, but, unbeknownst to Einstein, one could generate the field equations Einstein sought directly from Mie's ideas. Mie's work held the solution Einstein was struggling so hard to find, but Einstein would not be the one to figure that out.

Minkowski had died four years earlier, in 1909, regretting that he could not see the extension of relativity that he predicted. But his dearest friend, Hilbert, had maintained an interest in the subject and thought about the field equations that would sit at the heart of the theory. There is controversy in some quarters about whether Einstein or Hilbert deserves credit for the discovery. The two spoke and corresponded about their mutual interest, and Einstein was the first to publish, but Hilbert's version emerged at almost the same exact time, possibly weeks earlier. Although the two were in contact and therefore not working fully independently, they did take different routes to the same equations. Einstein started with physical intuitions, while Hilbert derived the theory mathematically from Mie's work. There is little doubt that Hilbert did find the field equations for the general theory, but he was very clear in giving Einstein credit. The famous dig at Einstein's mathematical abilities comes with a backhanded compliment and priority for discovery. "Every boy in the streets of Gottingen understands more about four-dimensional geometry than Einstein. Yet, in spite of that, Einstein did the work and not the mathematicians."[22]

With help from Grossmann, Hilbert, and a number of others, by 1915 Einstein had developed the field equations that would sit at the heart of the general theory of relativity. But just when it seemed the theory might be complete, he found that he had run into what might be an intractable concern, a conundrum that might undermine the entire project—Einstein called it the "hole problem."

The key to the general theory of relativity is to completely reimagine gravitation. Since Newton, we had thought of gravity as a force in space. Space, according to Newton, was fixed and immovable, an inert underpinning to the world. Space was a thing in itself, he thought, but it was a thing that never interacted with anything else. It was flat and regular. The things in it, however, did affect each other. When there was contact among the things, the interactions were governed by Newton's laws of motion, and when not in contact, gravitation caused each and every object to attract each other. Gravitation is a force between the objects and the vast reaches across space, but it is not derived from space. As such, the theory contained a space that obeyed Euclid's geometry and had an ever-changing gravitational field contained within it. The gravitational field is just a set of values for the gravitational pull at every point in space; that is, for every spot there is a number and an arrow—the number tells you how hard gravity would pull you and the arrow tells you in what direction.

One problem for Newton's theory was that these arrows and numbers changed instantaneously with the movement of objects in space. As things traveled through space, the gravitational field adjusted itself timelessly. But in relativity, there was a maximum velocity for signals. No information could travel faster than the speed of light, yet for Newton, gravitational news of an object's motion did exactly that. If Einstein was right, then gravity had to adjust itself as a wave spreading out through space with a finite velocity.

But it was not only space that would be affected. Minkowski had shown Einstein that space and time could not be separated, and so Einstein realized early on that time dilation effects would be seen in strange ways in gravitational fields. In the paper he wrote for Stark's journal, he pointed out that clocks should run slower in a stronger gravitational field–for example, a clock on the sun should run slower than a clock on the earth. Gravitation would affect both space and time.

Gravity is unique among physical forces. Consider electricity. Like charges repel and unlike charges attract. If you have two electrically charged particles, p_1 and p_2, then they will attract or repel each other based on their charges and the distance between them. If you keep them separated at the same distance, but double the electrical charge on p_1, then you also double the acceleration that it experiences from the force between p_1 and p_2. Nothing strange there: more charge, more acceleration. But now take two particles with masses m_1 and m_2. They attract each other with a gravitational force based on their masses (their gravitational charge) and the distance between them. Now double the mass of m_1. You would think that it would likewise double the acceleration from the gravitational force between the two particles. But it doesn't. An eight-pound bowling ball and a sixteen-pound bowling ball fall at the same rate. Only gravitation behaves this way. Gravity differs from all other forces in the universe.

Einstein's "happiest thought" led him to the idea that gravitation and acceleration were different ways of looking at the same thing. But acceleration is tied to how one views space. Think back to the elevator example. If you are in an elevator with a bathroom scale, then you can tell if you are stationary or in free fall. Step on the scale, and if it shows your weight, then you are at rest. If it reads zero, then you know that you and the scale are falling at the same rate. Now suppose you took the scale into a rocket in space, far removed from any mass. Since there are no masses around, there would be no gravitational attraction to pull you down and you would be weightless: the scale would read zero. But if you fired the rocket's engines and accelerated just the right amount, the scale would read exactly your weight on earth. The idea is that if you were in a small metal room with nothing but your scale, you could not tell if you were accelerating or in a gravitational field. If the scale reads your earth weight, you could be at rest in the earth's gravitational field or accelerating where there is no gravity. If the scale reads zero, you could be accelerating in a gravitational field or at rest where there is no gravity. There is no way to tell the difference, so the two must be different senses of the same physical state.

But there does seem to be one way you could tell. Poke a hole in the side of one wall. Light will enter that hole and shine on the wall across from it. If you are at rest, then the spot of light from the hole will shine on the opposite wall at the same height because light moves along the shortest path between points. But if you are accelerating upward, then the spot of light will appear lower than the hole because the wall will have moved upward with the rest of the room during the time it took the light to travel across the room. An asymmetry seemingly undermines Einstein's equivalence—but only if we assume that light always moves in a straight line. In fact, light moves along the shortest path in space, which is a straight line only if the space is flat. If the space is curved, then the shortest path will also curve. Consider a globe. If you want to move along the globe from one place on the equator to another place on the equator, the shortest path is along the equator—but the equator is not straight, it is a curve. So if the dot of light is at the same height as the hole, it might mean that you are accelerating where there is no gravity, or it might mean that the light's path is bent by the effect of gravity. Since light travels along the shortest path from one point to another, this would mean that the effect of gravity is to bend the space through which the light traveled.

This is where Einstein had to seek Grossmann's help. The shape of space was determined by the distribution of matter and energy in the universe, and the way that matter and energy moved through space was determined by its shape. This curved shape and the force experienced in it could no longer be represented by an arrow and a number at each point; rather, a number was required for each different direction at every point. How to handle calculations with these arrays of numbers and how to account for the variable shape of space were what Grossmann taught Einstein. But Einstein needed to take these tools, this new language, and set out the equations that could predict real-life observable *phenomena* in those terms. This took years, but Einstein finally did it.

The gravitational waves he predicted could not yet be observed. Nor could the shift in the color of light in a gravitational field that followed from the slowing of the clocks. The bending of light might be observed, but it would have to involve light traveling very close to a massive object, something as heavy as a star. Such a phenomenon might be viewed during a solar eclipse, and plans were set up to try to make the requisite observations during the eclipse of 1914. But the start of the First World War made that impossible.

There was one other possibility for an observable result from the theory that was not predicted by Newton, and it involved something that was well known—the advance of the perihelion of Mercury. Aristotle held that all heavenly bodies move in perfect circles, but observation quickly showed that we could not account for the actual trajectories of the stars and planets in terms of uniform circular motion. Ptolemy added epicycles, circles on circles, and eclectics, off-centered circles, to account for the differences and made the whole system work. Nicolaus Copernicus had argued that we needed to put the sun and not the earth at the center of the model, but this still required epicycles. Johannes Kepler finally did away with the epicycles, showing the simplest path to be not circular but oval. Planets move around the sun in ellipses. Newton's laws of motion explained this elliptical motion; indeed, initially this was one of the strongest pieces of evidence for Newton's theory.

But some planets had orbits that were not quite elliptical in the way they were expected to be. For example, the relative size and distance from the sun of the planet Uranus, a late discovery, could be determined using Newton's theory, but its orbit deviated from the expected route. Even taking into account that planets were attracted not only by the sun, but by all other masses as well, including other planets, Uranus still did not behave as scientists thought it should. Perhaps there was an additional influence, such as another massive planet close by.

Based on its path, scientists scrambled to determine how big and where such a new planet would have to be, and that work led to the discovery of Neptune.

But Uranus was not the only planet with an anomalous orbit. Mercury's orbit was also odd—like the curve drawn using the old Spirograph toy, it was an ellipse that never closed. When Mercury completed an orbit around the sun, it did not return to the same place; its orbit advanced slightly each trip around. If we picture Mercury beginning its orbit at the twelve o'clock position, the next orbit would begin at the twelve o'clock plus three seconds mark, with each subsequent orbit starting another three seconds further around. Scientists proposed that this perturbation was the effect of another mass; maybe another planet was situated inside Mercury's orbit, a planet we had never seen before because it was so close to the sun that the sun's light obscured it. Astronomers researched ways that such a planet, dubbed Vulcan, could be observed, but all attempts failed. Without such a planet, Mercury's strange orbit was left unexplained.

Einstein had thought from early on that this effect could not be accounted for by Newton and that it would be the first great test for his new gravitational theory. With each new formulation of the general theory of relativity he proposed, Einstein calculated what his theory would predict in terms of the advance of Mercury's orbit. Each time it failed to give the proper amount, and he knew something was amiss. But in 1915, he used his latest version, and the result was spot-on. Einstein said that he suffered palpitations as if something inside of him had burst, he was filled with such immense joy that he could not focus on physics for several days. He knew he had it. The general theory of relativity was within reach. Einstein had a working theory. He had equations that did what he wanted. On one side of the equation was the distribution of mass and energy. On the other side was the geometry of space. Put them together and they accounted for everything Newton's old theory did, and for some anomalies it did not. After years of struggling and having to train himself in new mathematical techniques, he had it.

And then came the "hole problem." If the theory of relativity was successful, you should be able to input the location and amount of all mass and energy in the universe and get out a unique and complete description of the curvature of space-time at every single point—that is, you should be able to absolutely determine the way the gravitational pull would affect an object if it were situated anywhere in the universe. But consider a hole, a small region of space out in the middle of nowhere, a volume of empty space with no mass or energy contained within it. We could take the values for the gravitational pull inside the hole and twist them, map them on to points nearby so that every point inside of the hole had a new gravitational value. As long as the values stay the same outside of the hole, everything would work exactly the same in the universe. Remember that the hole is empty (that's what makes it a hole), so if one were to take a God's-eye view of the universe, things would look identical—every event in the universe that has the original calculated values for the points in the hole would happen exactly the same way as in the universe with the twisted values in the hole. But while the universe worked the same, the underlying space-time geometry was different, and it was the geometry that was supposed to make things happen. If there were two possible geometries and the theory gave one and only one, then the theory could not be complete. Which universe did we really live in? The theory could not determine it. This meant that Einstein's theory of gravitation had the same sort of problem that had plagued Newton with his theory of absolute space and Maxwell with the luminiferous aether. Einstein's theory had an unobservable substratum now, too, and along with it the very problem he thought he had solved with his special theory of relativity. All that work, and he ended up in the same old pickle.

Einstein was flummoxed. He went back over the physical reasoning and the mathematics. Everything seemed in order. What could be the problem? Then it struck him. Einstein had to clearly state the line of reasoning he had been making for a decade. It was a version of what the seventeenth-century philosopher Gottfried Leibniz called the principle of the identity of indiscernibles—if there is no possible way to tell the difference between two things, then they are not in fact different, they are the same thing. Just as with the case of the magnet and the coil, and the principle of equivalence with the person in the elevator, so too with the space with the hole: the universes with the different field values for the gravitational pull inside the hole are just different descriptions of the same thing. The field values are mere tools; what is real is the set of all events, all things and their interactions. The value for the curvature of a point in space is meaningful only if it makes a difference. We defined the hole in a way that guaranteed that it made no difference; that is what made it a hole. The problem, Einstein realized, was not with the physics at all, but rather with the meaning assigned to the terms in the physical theory. It was a philosophical problem, one that could be solved by adopting the view that Einstein had admired so much in the writings of Ernst Mach, whom he read back in Bern with his friends in the Olympia Academy. According to this view we should take observability as a basis for assigning something the status of a real thing. It was this insight from Mach that led Einstein to the theory of special relativity, and now that he realized he had fallen into the same trap Mach had warned about, he was able to reinterpret the general theory of relativity so that the hole problem was no longer a problem.[23]

In 1916, Einstein published his theory in an article titled "The Foundation of the General Theory of Relativity," which appeared in issue number 7 of that year's *Annalen der Physik*. It is a triumph of elegance and imagination. It is a radical revision of our understanding of the nature of the universe itself. Isaac Newton's theory of gravitation, space, time, and motion had dominated physics for three hundred years, standing as the single greatest achievement in the history of science. Here was its successor.

4

Two Wars

WITHIN MONTHS OF Einstein's move to Berlin, he found himself immersed in two wars. One was geopolitical and pitted Britain, Russia, and France against Germany and the Austro-Hungarian Empire, while the other was personal and was fought between Mileva on one side and Einstein and Elsa on the other. Switzerland may have been neutral territory in the Great War, but it was the place where Mileva dug in and fought bitterly for her marriage. Both of these wars had casualties and made life difficult for combatants and innocent victims alike.

When Einstein left Germany as a high school student and renounced his citizenship, he swore to himself that he would never go back. With the offer from Planck and Nernst, he relented. But in returning he promised himself he would keep his mouth shut and not speak out against German policy. That promise would be broken as well.

The frenzy around the start of World War I reminded Einstein of everything he hated about the Germany of his youth. He had despised the militarism and nationalism that were on display with the regular military parades and present in the classrooms. He loathed the aggressiveness and arrogance that permeated the culture. At the start of the war, Einstein saw his worst fears regarding the German character coming true. Not only was there a sense that offensive military adventures were justified in the name of German ascendance, but there was near universal support for them.

The assassination of Archduke Franz Ferdinand in Sarajevo in June of 1914 tipped the first diplomatic domino, which, because of treaty-bound alliances, led the majority of European nations into war once the Austro-Hungarian Empire declared war on Serbia. German Kaiser Wilhelm II found himself at war against his cousins King George V of Britain and Tsar Nicholas II of Russia.

Born with a withered left arm that he usually rested on a sword worn at his waist, Wilhelm sought to appear intimidating instead of handicapped. He surrounded himself with sycophants and was well known for being both capricious in his decision making and obstinate once his decision had been made. He had to be the center of attention in every room, even when the room contained other royalty. Indeed, his antics made him deeply unpopular with members of his extended family, especially with his cousins in the British royal family (Queen Victoria was his grandmother on his mother's side), and as a result anything relating to England received an explosive emotional response from Wilhelm.

Grandson of the first Kaiser Wilhelm and son of the second Kaiser, Friedrich III, who died of cancer shortly after becoming ruler, Wilhelm II resented Chancellor Otto von Bismarck, whom he saw as reducing his family to figureheads and illegitimately usurping the power to rule. Sacking Bismarck, Wilhelm sought to centralize control in the position of the

Kaiser. Assured of his own greatness and thereby that of Germany, he sought to advance national interests in terms of both military strength and nonmilitary fields of culture, arts, science, and industry. Germany was to become the dominant world power, and it was his leadership that would put it there. Wilhelm's sense of ascendance infected the culture. At war with its neighbors, Germany hatched a plan to defeat the French, who not only were their enemies in this conflict, but had been considered despicable since Napoleon's conquest of the Germanic territories when they had belonged to the Holy Roman Empire. But to occupy French soil, they would need to move through Belgium, which had declared itself neutral.

The Germans invaded Belgium on August 4, 1914, arguing that the move was preemptively defensive. Small, lightly armed Belgium never posed any threat to Germany but, Wilhelm contended, was a looming peril that had to be addressed. The fight was more difficult than planners had originally thought, however, and resulted in massacres of Belgian civilians, shocking the world.

Germany's opponents seized on the Belgium atrocities, propagandizing the slaughters and demonizing the Germans. One angle that the propagandists took was to contend that there were "two Germanys." "The people of western Europe asked with astonishment: 'How can the German people, whose music we love and whose science we admire, be capable of such unlawfulness and such atrocities?' Partly for propagandist reasons there was invented the story of the 'two Germanys,' the Germany of Goethe and the Germany of Bismarck."[1] The German intellectuals were not to be blamed for, or even associated with, the horrific events that had occurred.

This line may have been soothing to the Germanophiles on the rest of the Continent but was strictly false in terms of facts on the ground. Support for the war was nearly universal in Germany, cutting across political, social, cultural, and economic lines. The Kaiser was roundly lauded when at the start of the war he proclaimed before the Reichstag, "I no longer see any parties, I now see only Germans."

German intellectuals resented their international colleagues trying to portray them as alienated from the war that would establish Germany as the world's great power. To put an end to the "Two Germanys" notion, they produced "The Manifesto of the Ninety-Three," so-called for the ninety-three signatories representing every corner of the German intellectual world. The Manifesto contended that the invasion of Belgium was purely a matter of defense, that no atrocities were committed by German soldiers, and that one cannot separate German militarism from the rest of German culture. It was signed by fourteen scientists who had already won, or who would later win, the Nobel Prize, and included Einstein's friends and colleagues: Planck, Nernst, Lenard, Fritz Haber, Paul Ehrlich, Wilhelm Ostwald, Wilhelm Roentgen, and Wilhelm Wien. These were the people who influenced Einstein's work and with whom he corresponded as a young physicist first making contact with the world of academic physics. They had championed his work when it was still deeply controversial, and they met with him regularly now that he was back in Germany.

Einstein was sufficiently alienated from his colleagues that this political difference did not bleed over into his working life. Einstein did not participate in the faculty meetings and internal politics of the university or the Prussian Academy. He remained aloof from the pettiness of the intellectual community, only talking shop with his fellow scientists. But he could not remain neutral in this fight—after all, look at what had happened to Belgium when it tried to maintain its neutrality. So, when his friend and fellow pacifist Georg Friedrich Nicolai drafted a counterdocument titled "Manifesto to the Europeans," which eschewed nationalism as harmful to humanity, culture, and intellectual progress, Einstein was one of four intellectuals who signed it. "We recognize the need for international relations which will necessarily move in the direction of a universal, world-wide civilization," the manifesto declared, going on to argue that Europeans needed to seek peaceful means of settling any diplomatic differences in the name of unity and human progress. At a time of fervent nationalistic furor, Einstein was publicly calling for an end to nationalism. At a time of fervent clamoring for military conquest, Einstein was publicly calling for an end to war. At a time when lines were being drawn between us and them, Einstein was publicly calling for there to be no them, that they are us. The war was a tragedy in Einstein's mind, the horrible end result of the cultural excess he came to loath as a child and against which he felt the need to speak out.

Along with Nicolai, he helped create the New Fatherland League in November of 1914, a loose group of internationalist and pacifistic-minded intellectuals and professionals. Ultimately they sought a "United States of Europe" where the scourge of nationalism would no longer lead to war. The organization was raided and shut down by the government, but not before the group drafted and distributed literature on their ideas, often in places where they were not allowed, and worked on behalf of fellow pacifists who had found themselves in legal difficulties. Einstein eagerly engaged in activities that would have led to arrest had he been caught. In the words of one of his co-conspirators, "Einstein took an almost impish pleasure in pulling a fast one on the military authorities. Each time we succeeded in smuggling letters or books into a prison, he would laugh aloud in amusement."[2]

While he opposed the hostilities on the battlefields of Europe, he was an active combatant in the hostilities between the women in his life. The Great War began in the Austro-Hungarian Empire between the German and Slavic factions. The war among the Einsteins pitted Mileva against Elsa.

Deeply hurt by Einstein's disaffection, Mileva became deeply jealous of anyone who received Einstein's attention, occasionally lashing out in a way that was problematic for all involved. So when he rekindled his relationship with his cousin in Berlin, Mileva violently disapproved. Mileva's accusations concerning her flirtatious husband were sometimes misguided, but with respect to Elsa they were correct.

Einstein found himself living a bachelor's life in Berlin once Mileva had taken the children back to Switzerland. Working on the final difficult elements of the general theory of relativity, Einstein focused on his work to the exclusion of his health, hygiene, and well-being. Elsa took it upon herself to take over the details of day-to-day life that Einstein neglected—seemingly trivial aspects such as eating and grooming that he thought unnecessary as he grappled with the deeper mysteries of the universe. She was divorced, with two older daughters, and accustomed taking care of others, so Einstein would fill that gap for her, a situation that more than pleased Einstein, who not only needed the help but enjoyed the distant but caring admiration.

In Elsa, who had a simple charm and a sunny disposition, Einstein saw the opposite of Mileva. Mileva was an intellectual who sought to be a gender-barrier-breaking pioneer and Einstein's intellectual partner, considering herself to be his equal. Elsa put her cousin on a pedestal, never invading his work but instead caring for his more basic needs. She would baby Einstein, admonishing him when it was time to eat and forcing him to dine when he would initially wave away her home-cooked meals. Elsa saw herself not as essential to the production of his work, but as more of a business manager, dealing with the emerging public side of Einstein, something he cared little for. She was a Swabian in Prussia, a country bumpkin in the big city, and that brought with it a sense of inferiority that could be overcome by her association with her cousin. If people were clamoring to have face time, interviews, and collaboration with him, then she would have power by being his gatekeeper. It was a post he needed filled and which she performed to his liking. The relationship suited them both.

Mileva had taken the children back to Zurich and was estranged from Einstein, but she harbored the hope that they would be reconciled, that he might return to their Switzerland, and that the issues between them could be resolved. She despised Elsa, demanding that she not be present should Einstein visit the boys. For her part, Elsa made nasty claims about Mileva, such as accusing her of wanting to harm Einstein. Both accused the other of trying to control him and being motivated by greed. The relationship between Einstein and Mileva became increasingly frosty, and whether through direct correspondence or through intermediaries, the vitriolic tone between them was unconstrained.

The stress of the times and the internal struggles took their toll on the entire family. Eduard, the younger son, became increasingly ill, requiring extended stays in a sanitarium. Mileva suffered from scrofulosis, a version of tuberculosis, and for a while was bedridden and unable to care for Hans Albert, sending him to live with family friends. At times, she, too, would have to spend time at a sanitarium. Einstein also became ill, suffering from gallstones, a liver problem, stomach issues, and a duodenal ulcer, conditions that would plague him for four years and that would continue to flare up for the rest of his life. Elsa, witnessing Einstein's poor health, moved him closer to her so that she could nurse him more effectively.

The situation with Mileva continued to deteriorate, and Elsa took up more and more space in Einstein's life and pressed him to marry her. The current arrangement suited Einstein just fine, and so he had no desire to remarry or to pursue a divorce with Mileva. But Elsa had two daughters who were approaching marrying age, and the whispers about their mother's relationship were harming them socially. Elsa's family pressed Einstein, and ultimately he relented. He asked Mileva for a divorce, but she declined. To try to secure the divorce without her agreement would be impossible, because evidence or admission of an affair was required and, while Einstein had suspected that Mileva had been unfaithful, he had no proof. A protracted negotiation followed. Not only did Einstein have no leverage in the bargaining, but because of the war he would have difficulty converting his German currency into Swiss francs to meet the financial terms of a settlement. Mileva did not want a divorce and certainly would not settle for one without sufficient provisions for herself and the boys.

Once the war ended, negotiations between the Allies and the newly formed Weimar government would proceed increasingly well, though every German was facing a difficult financial situation. The reparations brought on hyperinflation in Germany, undermining the wealth and savings of the nation. Einstein, with everyone else, was in need of foreign hard currency. He would get it, with help from the British.

Arthur Eddington was a rising star in British science. He had been the chief assistant of the Astronomer Royal at the Royal Observatory in Greenwich before moving to Cambridge and occupying the chair of the Plumian Professor of Astronomy, at which time he was elected to the Royal Society. He was becoming a major player in the astronomical world. But there was a problem: as a Quaker, and therefore a pacifist, he objected to the war. Like Einstein, he was an internationalist who called for worldwide collaboration. He was an embarrassment to the government during a time of enforced patriotism. He needed to be kept out of the public eye. He needed to go away. And so away he went.

Astronomers had looked through telescopes and collected data since the days of Galileo, but the introduction of photography had revolutionized astronomy. Instead of the human eye engaged in direct observation, photographic plates could capture images from telescopes and the resulting pictures could be analyzed in much greater depth. Eddington had been one of the pioneers in developing the mathematical means of interpreting the data on these photographic plates.

Einstein had grown friendly with the scientists at the University of Leiden in Holland, where H. A. Lorentz—Einstein's hero—was the patriarch. Einstein would visit the university to discuss physics and sit at Lorentz's feet. Lorentz and Einstein had grown close in these years; Lorentz represented a father figure to Einstein, and he felt the sort of pride in Einstein's progress that one would experience toward a son. When Einstein received the offprints for his paper on general relativity, he sent them to several colleagues in Leyden, including one to the astronomer Willem DeSitter, who passed it on to Eddington. Because of the war, German publications were not available to British scientists, but DeSitter knew that Eddington would be quite interested in Einstein's work. Indeed, he was.

So, with the upcoming eclipse of 1919, Eddington realized he had a chance to contribute to the project. The full eclipse would be seen close to the equator, and Eddington would oversee teams sent to Brazil and to Principe, an island off the coast of West Africa, to capture the event through telescopes on film from two locations and thereby test Einstein's theory. Eddington would accompany the team headed to Principe. It would be important science, and it would get Eddington out of England in a fashion that served the crown. Travel to such distant and undeveloped places was difficult and time-consuming, especially with the threat of German U-boats shutting off more direct routes. It was a project that would take Eddington away for a good long time.

Fortunately, the war ended just as travel began, and the teams arrived safely and in time for the long process of setting up, testing, and calibrating their equipment. It had taken months for the teams to make their way to their appointed locations. Camping out under primitive conditions while trying to set up and care for highly sensitive scientific equipment had been incredibly difficult. Illness and supply problems also plagued the mission. The day of the eclipse finally arrived, and the scientific teams worked frantically to make sure all systems were prepared to photograph the sky. But in Principe the day dawned rainy, with the sky covered with thick gray clouds.

Fortunately, the god of physics smiled. The rain ended and the clouds parted just in time for the eclipse, and photographic plates were made from both locations. Eddington returned to England and set about analyzing the results. The images on the plates from the team from Brazil were fuzzy, and although the resolution was not good enough to allow for a confident assessment, the results seemed to indicate that Einstein was wrong, that the Newtonian result was correct. But the plates from Africa were much sharper and the data more exact. These results showed Einstein's prediction about the bending of light to be virtually spot-on. When the data from the two were combined, Eddington concluded that Einstein was the victor. The theory of general relativity, he reported, had been experimentally confirmed.

Before the result reached the outside world, a telegram was sent to inform Einstein. He read it and handed it unceremoniously to Ilsa Schneider, a graduate student with whom he was talking, saying, "Perhaps this will interest you." She quickly realized it was monumental news for the whole of mankind and was stunned that Einstein seemingly remained unaffected. Baffled, she asked why he was not overjoyed. He responded that, eclipse or not, he was absolute in his confidence in the theory. She inquired what he would have thought had the results of the observation gone the other way and vindicated Newton's theory. Characteristically self-assured, he told her, "I'd have pity for the dear Lord . . . the theory is right."[3]

The result made headlines worldwide. In the aftermath of the unbearable death and destruction from the Great War, here was good news, interesting and strange news that the universe was an exciting, dynamic place and that we would need to understand it in a completely new way. It caught the fancy of a world tired of thinking about mankind as barbarians and eager to celebrate its creativity and insight. And at the center of it was this curious, unkempt, wisecracking figure who seemed to stand for a different side of humanity.

At first, the petty nationalistic resentment of the age was mapped onto the scientific revolution that was occurring. Could Einstein, a German, have really replaced the accomplishments of Isaac Newton, the pride of England? Did Eddington, a Brit, really just hand the throne of physics over to the enemy? But once Einstein's antinationalist commitments and internationalist sympathies became evident, such talk disappeared and Einstein was hailed as a hero across the globe, held up as the model of the new German, the good German, the rehabilitated German.

Einstein was an attractive figure for the world press. Visually, he was as bizarre as the theory he was making famous. He was always quick with a witty quotation that made for great stories. Elsa worked hard to keep Einstein in front of reporters, but not too much, finding a balance between exposure and overexposure. He was scarce enough to remain desirable, but accessible enough to warrant seeking out.

For Einstein, this status as an international celebrity was a double-edged sword. On the one hand, it was the quintessence of superficiality and, as with all other cultural vacuity, he disliked it and mocked it. On the other hand, he had his playful side and could play it for the joke it was, and it was a joke that paid handsomely. Einstein starting receiving offers to speak for large sums of money paid in hard currency.

He overplayed his hand at first, requesting absurdly large speaking fees for transatlantic travel (he was not entirely enthused about undertaking such trips, so the rejection did not concern him too much), but he eventually figured out the game and learned to play it well. The fees could be sent directly to Mileva for her and the children without having to go through Germany and thus avoid devaluation. This helped to ease tensions somewhat and gave Einstein a place from which to negotiate an end to the marriage. The peace accord that ended the geopolitical war in Europe coincided with the truce that ended the personal war among the Einsteins. The ultimate bargaining chip that Einstein played was his promise to give Mileva the entirety of the money he would be awarded when he won the Nobel Prize—not *if* he won it, but when. He was confident that it would happen. He had been nominated repeatedly, but each time someone else received the notification. It would come, he was sure, and the amount of the award was so significant that Mileva, who also had little doubt that it would in fact happen, agreed.

Einstein filled out the paperwork in January of 1919 and traveled back to Switzerland, where he testified that he had been living with his cousin while still married to Mileva, and the divorce was granted. As part of the agreement, he was forbidden to remarry for a period of two years. But since Einstein was divorced in Switzerland and was to be remarried in Germany where the divorce terms would not be enforced, he waited only a couple of months before making it official and Elsa became Mrs. Albert Einstein. The fact was, however, that "Einstein" was Elsa's maiden name, and for appearances she had gone back to using it for quite a while beforehand.

Einstein had begun to travel abroad with great frequency, partly because important scientists with whom Einstein wanted to speak were inviting him to visit. But in addition, the wave of nationalism that arose during the war and continued thereafter grated on Einstein's nerves. He considered it a time of cultural insanity and only by leaving Germany did he feel he could shed his "muzzle." He was frequently in Switzerland to visit the children and to lecture at both the University of Zurich and the ETH. Indeed, the two schools worked together with the government bureaucracy to create a joint position designed especially to lure Einstein back. But Zurich belonged to Mileva and he now belonged to Elsa, so he turned it down.

Einstein was lecturing regularly about Eddington's results, which focused his mind on the large-scale astronomical ramifications of the general theory. Einstein came to a shocking realization about what the theory said about the universe—it was not stable. If his field equations were correct, then it would mean that, but for an instant, the universe would always be expanding or contracting. That couldn't be right; an unstable universe just would not do. This needed a fix, and that meant that the field equations, which he thought were right as far as they went, did not go far enough. The theory was incomplete and required an additional term to make sure the universe remained static. In a paper published in 1917 by the Prussian Academy of Science called "Cosmological Considerations of the General Theory of Relativity," he added the *cosmological constant* term to his famous equations, and its only function was to counteract the expansion or contraction of the universe that was an accidental consequence demanded by the rest of the theory.

It would be, by his own admission, Einstein's greatest mistake. For Einstein, the theory was true because it was elegant. The poet's correlation between truth and beauty held for Einstein as well. If a theory needed patches or additional clunky theoretical baggage such as an absolute space or a luminiferous aether, then that showed that there was a problem with the theory, a problem that needed correcting—generally with a new theory. It was by eliminating such superfluous elements that Einstein had made his great discoveries. Yet here was Einstein committing precisely the sin he had shown to undermine centuries of physical thought, adding unnecessary elements into a theory to make it accord with preexisting biases concerning how things were believed to be.

His monkeying around was eventually discredited when Edwin Hubble discovered that the universe is indeed expanding. He was able to show this based on the shift in frequency observed in starlight. This is the result of the Doppler shift and shows that virtually all of the galaxies we see are moving away from us, and the farther away a galaxy is, the more likely it is to be moving away faster. The best explanation for this is an expanding universe. But if it is getting bigger now, it must have been smaller in the past, and something had to cause the expansion. This discovery gave rise to the Big Bang theory of George Gamow, and it led Einstein to embarrassingly admit that he should have allowed himself to be led by the science, and not have tried to shoehorn the theory into a picture colored by prescientific presuppositions.

Although his work found more and more widespread acceptance, the most prestigious award, the international symbol of one's participation in the grand march of the history of science, the Nobel Prize, continued to elude him.[4] His name was put forward for ten of the twelve years between 1910 and 1922, but every time he was denied. This was partly because the Nobel committee regarded science as necessarily experimental, so the award should be given only to those who make discoveries by observing new things. The traditional thinking was that physics began with the world, where careful and crafty scientists would painstakingly document surprising phenomena, to have them later explained. The rise of theoretical physics, where ideas came first, then mathematical formulation, and empirical evidence was sought only after the fact, was new and not appreciated by the older generation of scientists. By their view Einstein didn't discover anything, he just told other people where to look—and they had to do the real science in finding it. Science requires a lab coat, not just a pad and pencil.

But there were other grounds as well. The award is given based on a vote of the entire Swedish Academy on candidate names that come out of discipline-specific committees. The larger body is not bound to follow the advice of any committee, and it is not unheard of for the committees to be overruled. The Academy was composed of older scientists, doctors, and engineers who saw Newton's work as the basis of the scientific endeavor. They were among those who thought that at best the jury was still out on relativity. Ignoring any supporting evidence, they would continually cite every observation that seemed not to support Einstein to show that his view was far from confirmed.

Each year, important physicists from across Europe would write letters in support of Einstein's nomination, with a growing number arriving from increasingly better-known figures in the scientific world. Eventually the chorus from physicists became overwhelming. Eddington compared Einstein to Newton. Wilhelm Ostwald, the 1909 Nobel laureate in chemistry who was the last great holdout against atomism and whose mind was finally changed by Einstein's work, compared Einstein's breakthroughs to those of Copernicus and Darwin. Planck lobbied hard. Marcel Brillouin, a notable French physicist, chided the Academy in his supporting letter, asking the members to imagine the damage that would be done to the Nobel itself if, fifty years later, people looked at the list of recipients in physics and did not find Einstein's name.

The pressure became such that in 1921 the Academy decided to form a special task force to examine the Einstein question. Among all of the achievements cited for his deserving the prize, the two put forward most often were the theory of relativity and his work on the photoelectric effect. The Academy assigned a senior member to examine each claim. Svante Arrhenius, the Swedish chemist who won the Nobel in 1903, investigated Einstein's work on the photoelectric effect. Arrhenius argued that Einstein's work was solid but argued against awarding the Nobel Prize for it. Einstein's use of the notion of light quanta was effective and did contribute to the development of quantum mechanics, but the Nobel in 1918 went to Planck for virtually the same thing. If the award was to be given for the photoelectric effect, then it should not be Einstein, who just came up with a theoretical explanation for it, but rather Philipp Lenard who should be recognized for the discovery, and he had already received the award in 1905.

The relativity report was written by Allvar Gullstrand, who received the Nobel Prize in medicine in 1911 for his work on visual perception. Gullstrand was an ophthalmologist, but his research combined physics, biology, and medicine to give the most rigorously developed account to date of the working of the eye. He was widely respected and, despite being a medical doctor, was held to have the requisite mathematical and physical background to assess Einstein's work. He was, however, deeply antipathetic toward relativity and cited objection after objection as he argued that the theory of relativity was a passing fad that had caught the public's fancy but was ultimately destined to be forgotten. He concluded that it would harm the Academy and the Prize to award it on the basis of popularity and not for legitimate scientific merit.

Based on these reports, the Academy rejected Einstein again in 1921. Indeed, no Nobel Prize in physics was awarded at all that year.

But the pressure did not ease. Physicists continued in more and more strident tones to demand that Einstein be recognized with the Nobel. Planck proposed that the Danish physicist Niels Bohr receive the prize for his contributions to quantum mechanics, but that since Einstein's ideas set the table for Bohr, he should be granted the previous year's unawarded award.

Once again, Gullstrand was asked to report on the state of confirmation of the theory of relativity, and his report was again strikingly negative. The photoelectric effect report was reassigned to Carl Wilhelm Oseen, a younger colleague of Gullstrand's at the University of Uppsala and a theoretical physicist. The fact that Oseen was assigned the task is important because Oseen had been among those who nominated Einstein. He was clearly on Einstein's side in the matter, and his report was much stronger than that of the previous year. But Gullstrand remained firm: relativity was not Nobel-worthy.

Oseen knew this would be the case, and, instead of nominating Einstein for both relativity and the photoelectric effect, he followed Planck's lead and chose to focus solely on the latter. This created a political opening that might be exploited. Suppose it was made explicit that the theory of relativity has absolutely nothing to do with the reasons for awarding Einstein the Nobel Prize, would that be enough to satisfy Gullstrand? In the end it was, and the Academy awarded Einstein the 1921 Nobel Prize for physics in 1922. The official notification to Einstein made perfectly clear that he was being celebrated for his contributions to quantum mechanics and that his work on the theories of relativity were in no way the source of his award.

Einstein was not bothered. The Nobel Prize for him was simply a matter of assuring his children's well-being. The formality and the pretention held little attraction for Einstein, who chose not to be present to receive it. Einstein had been invited to give a series of lectures in Japan during the time when the ceremony would be held in Oslo. Einstein was even warned by Arrhenius that news of the Nobel was coming and that arrangements for the Japanese trip should be altered. But Einstein ignored the suggestion and left anyway.

This created a diplomatic problem for the Academy. Generally, if a recipient is unavailable to receive the award, a dignitary from the winner's home nation accepts it on that person's behalf. But, in Einstein's case, which country? Playing on the long-standing tensions and mistrust between Germany and France, Einstein famously quipped about his circumstances that "if my theory of relativity is proven successful, Germany will claim me as a German and France will declare me a citizen of the world. Should my theory prove untrue, France will say that I am a German, and Germany will declare that I am a Jew."[5] This was precisely the situation. Einstein's passport was Swiss. The Swiss claimed him as theirs. But he was born in Germany and was currently working and living there. The Germans were not about to give up the honor. They contended that even though Einstein held a Swiss passport, once he accepted the job in Berlin—a government-sponsored position—he became a Reich citizen, since only citizens of Germany could be government officials by law. The Swiss and the Germans both wanted credit for Einstein's success. In the end, the Swiss relented and allowed a German representative to accept the award.

But where the Swiss may have caved on the matter, Einstein would not. He contended that he was Swiss and not German and that, when he returned from abroad and was given the award, it would be presented to him by a Swiss official in Germany and not by the Germans. This put the Swedes in a tough diplomatic position, but they decided that the best route was a third option, and Einstein was presented with his award directly by a Swedish minister.

5

The Worldwide Jewish Celebrity

Albert Einstein renounced his German citizenship as a teenager at the same time that he renounced his Judaism. In his mind, both nationalism and religiosity were symptoms of a shackled mind requiring unthinking loyalty to a structure built on authority. As the unpleasantness surrounding his Nobel Prize illustrated, he had resolved himself with respect to citizenship. He openly declared himself to be Swiss despite the fact that he had turned down a significant offer that would have brought him back to Zurich, choosing to remain in Berlin.

While his status as a Swiss citizen was initially acquired for pragmatic reasons, that being to secure a job, the case was different with religious affiliation. When filling out the paperwork for his position in Prague, Einstein had to specify a religion. The law would not allow him to leave it empty or declare himself an unbeliever, and eventually he put down "Mosaic."[1] Similarly, on his divorce forms, he wanted to write "dissenter" as his religion, but the clerk again filled in "Mosaic."[2] At a time when Einstein openly declared himself to be Swiss despite not living in Switzerland, he still did not consider himself Jewish.

The pride in his Swiss citizenship came in part from his desire to remain alienated from Germany because of cultural elements that disturbed him as a child. Those aspects of German society flared up between the World Wars in a way that also made him view his Jewishness in a new light. Being Swiss, for Einstein, was not about whether he lived in Switzerland; rather it meant having a certain stance toward the world, an approach he had seen in his Swiss "papa" Jost Winteler. To be a citizen did not mean to embrace nationalistic sentiment toward your home nation. You could love your country and still be a free thinker. The fall of the Weimar government and the rise of National Socialism would create the same sort of dichotomy for Einstein with respect to religion. Einstein could become Jewish again in his own mind without having to surrender the scientific worldview, the personal ethic, or the metaphysical foundations upon which he rested his physical theories. Being Jewish became, for Einstein, similar to being Swiss. It was not a religious proclamation but something else, something more subtle, an inalienable aspect of his being.

World War I was to be a watershed event according to the mythology German nationalists were creating. Compared with the other nations of western Europe, the united Germany was a young country and, in a sense, coming late to the table of geopolitical supremacy. The others had all had glory days as reigning regional or fully global powers. The French, the British, the Dutch, the Spanish, the Portuguese—each had been colonizers and at various times held military sway over their neighbors. But through the nineteenth century and into the twentieth, Germany was finding its legs. Its industrial, scientific, cultural, military, and technological contributions displayed Germany's status as an international leader. Intellectual systems, from Hegel's view of history to popular understandings of evolution, were portraying changes as instances of human progress. Things were getting better, people were getting smarter, societies were becoming more cultured. It could be interpreted, therefore, that instead of Germany being late to the game, all that came before was mere stage-setting for Germany's eventual appearance. History was creating the preconditions for the possibility of German emergence. This was the narrative they were creating for themselves, and it was used to justify actions such as those taken in the invasion of Belgium.

Victory in the Great War was presumed from the outset. Wartime propaganda and censorship deprived the people at home from a sense of the actual trajectory of the conflict. As such, the actual end of World War I left German society shocked. The story was supposed to end differently. It was Germany's turn. How could Germany have lost?

Some denied that it had. Losing a war means that you are invaded and a foreign power takes control. That is what happened when Prussia defeated the French in the Franco-Prussian War. The Prussians took Paris and remained until the government agreed to demands. But no one had occupied German land—not until the armistice was signed anyway. The Kaiser had abdicated, but that was an internal matter, not something forced by foreign design. The war had gone badly, yes; according to plan, no—but surely not lost, not in the usual sense.

Yet, there was a new government—not just a new Kaiser, but a new form of government, a parliamentary republic of the sort found in France and Britain. This was the government that not only agreed to the armistice, but also negotiated its terms, including not just crushing reparations but loss of territory, a neutered military, and foreign troops on German soil. It was—as even some of the victors thought—an unjust peace that smelled of revenge.

It was an armistice that many German conservatives considered not only humiliating, but unnecessary. General Erich Ludendorff contended that the German army under his control was, in fact, winning the war until the liberals, socialists, and communists overthrew the patriotic Kaiser and established the Weimar Republic, which failed to support the military and stabbed it and the entire Reich in the back with a cowardly surrender and negotiations that were at best inept, at worst treasonous. Given the censorship and the course of the war, ordinary people had no reason to disbelieve him. Indeed, for the nationalists who saw things going badly on all fronts, it was a line that fit nicely with their worldview: if Germany did lose the war, it was not because the German army was defeated, but because those on the political left lacked the will to finish the struggle that would have led to German dominance, the global superiority it was owed by the course of history. Everything that was wrong in Germany was the fault of the new government.

And who was the real power behind this government? Look at its supporters, look at its ministers, and look at its actions. The new government was led by the Catholic Center Party, the Social Democrats, and the newly formed German Democratic Party, which, unlike the socialists, stood for political and economic liberalism, running on slogans such as "Nothing will sway us from the middle of the road." Indeed, even though the socialist Social Democrats had received a greater proportion of representation in the Reichstag, the centrist German Democratic Party (abbreviated with its German name DDP), with only 18 percent of the delegates, led the government. The DDP had announced its creation with an open letter published in the widely read and liberal-leaning newspaper *Berliner Tageblatt* and signed by major German figures, including a significant number of well-known Jews, among them the paper's publisher and Albert Einstein. Despite the fact that Einstein was a socialist, he joined the DDP and was an ardent and visible supporter of the Weimar government. The support of Einstein and so many other prominent Jews led conservatives to tar the DDP as "the Jewish party," and the Weimar government as "the Jewish government."

This "Jewishness" was further entrenched in their minds when the government began rolling back long-standing anti-semitic laws. Not only were Jews seeing social advancement because of the government, but the government was being run by them, with Jews being placed in ministerial positions for the first time in history.

For a short while, the German economy boomed under Weimar. Foreign investment, much of it from the United States, created a brief period of prosperity. In it, Weimar culture blossomed in all its avant-garde strangeness. The war had arisen on intellectual soil that was already questioning the nature of humanity. The near-simultaneous revolutions of 1848 had led people to reexamine the foundations of social, political, economic, and moral structures that had gone largely unquestioned for centuries. Radical new possibilities of how we organize ourselves, how we understand ourselves, and how we conduct ourselves seemed to be live options. Did life itself have meaning with the old superstitions upon which we had based it stripped away?

And then there was the war. Europe, which considered itself the most sophisticated culture in the history of humanity, had reduced itself to naked and sustained barbarism. The death of so many in such horrible ways—mechanized death, mass chemical poisoning, bombs dropped from the air—had shown that the most advanced reverted to the most inhumane aspects of the self. "Who are we?" seemed a question that could not be avoided, and the answers would lead to a questioning of the foundation of all human endeavors.

Artists, musicians, and architects freed themselves from traditional strictures, creating works that seemed anything but artistic. Sigmund Freud's picture of the human mind seemed to undercut the traditional picture of what it was to be human. And then there was Einstein, who forced us to adopt a completely new and, to many outside of the scientific community, strange, mysterious, and impenetrable picture of what the universe itself is and how it behaves.

The culture split. On one side were those who saw the new as progressive. These deep doubts were liberating the spirit from old ways, which hampered human development and the ability to flourish. By this view, classical notions such as nationalism and religiosity were childish and representatives of the old, immature worldview that needed to be overthrown. To them, the horrors of the war were the result of the slavelike dedication to traditional ways. It was love of God, country, and king that got them into the mess. The war was evidence that we needed a new way of being in the world, and the advancement of science provided a model. Science stood as the template for generating rational beliefs, leading to new ways of seeing the world, and undermining the old capricious lines that divided people. National boundaries mean nothing to them. Scientists of every nation speak the same language, mathematics. They cooperate and advance humanity together. They openly question the basic concepts through which we make sense of the world, no matter how well entrenched, and reformulate them in a fashion that leads not to discord but to consensus and progress.

Albert Einstein was the symbol of this scientific cosmopolitanism. He was adored, inspiring poems and architecturally bizarre buildings.[3] His science, combined with his politics during the war, gave him the status of the wise elder statesman among young rebels. The fact that people did not understand his theory of relativity did not diminish his social capital; to the contrary, it increased it. By being the keeper of the mystery, he was considered the high priest of modernism. His dress and unruly hair were surrealist fashion. His quips to the media displayed the sort of contempt for the old that was part and parcel of its overthrow. Einstein seemed to embody the picture of human progress embraced by the modernists.

But others opposed this movement. Weimar culture, in their minds, rejected everything that was good and true, everything that had stood the test of time, everything that had given life meaning. They saw modernism as the pretentious navel-gazing of those who sought to undermine German exceptionalism. Germany was on its way to its rightful place as the dominant geopolitical force, and it had been stabbed in the back by these cowardly intellectuals who tried to hide their weakness behind a wall of false refinement. These modernists played meaningless games that they claimed to have real results. But in the end, the nationalists thought, all they did was to negate everything that was authentically German. They were not just creating, they were destroying, leaving in ruins that which should have been the most venerated.

The model for their wrongheadedness was science. Looking at the war, the nationalists saw science not as a template for moving forward, but as the tool to undermine humanity. The carnage of the Great War was so staggering, so many more brave and patriotic young men were killed than in wars past. Why? Science. It was the use of the new technologies that came from it that facilitated a new kind of war. Science dehumanizes, reducing us to mere machines, no longer the noble bearers of souls. The rejection of the old concepts and the old ways, the overthrow of the Kaiser by the socialists, the scale of the loss of life in the war, all of it is proof that modernism is not only wrong, but a mortal threat to the Fatherland.

And there stood Einstein, not just a scientist, but a political figure undermining support for the war with his articles and manifesto—holding the knife that stabbed the Fatherland and its brave fighting men in the back. After the war he used his fame to support the Weimar government, which the conservatives saw as doing to Germany what their enemies in the war had failed to do on the battlefield—destroy German culture. And, of course, he was Jewish.

The cultural divide was stark. The rhetoric became increasingly hyperbolic. And then the economy collapsed. The payment of reparations sapped government coffers, and the collapse of the American stock market dried up the flow of foreign currency. Forced to shoulder the burden of an extremely expensive war, and facing creditors who refused to renegotiate an unsustainable debt, Germany chose to print its way out. This caused a rapid devaluation of its currency and resulted in hyperinflation, which rendered the savings of most Germans increasingly worthless. Next came the massive job losses and a full economic implosion. The far left and far right looked at those in charge and laid the pain and suffering at their feet. They negotiated the armistice and the terms. They oversaw the nation as it spiraled into debt. They deserved the blame. The Weimar leaders were little men in gray suits who were incompetent at best, dangerous at worst. In the minds of both the far left and the far right, they had to go.

Einstein's association with the Weimar government caused a backlash against him at home, but his sudden worldwide fame and his objector status during the war made him a figure that was attractive to foreign countries. The French and British turned away anyone or anything German from their soil. But Einstein's stance on the war, his dedication to internationalism, and the striking nature of his scientific work led to his being able to procure invitations to speak where other German scientists could not. This meant that Einstein could be a unique cultural ambassador for Weimar and a force for reconciliation and worldwide peace. It would be an opportunity he would not pass up.

In seizing the chance to speak in political settings, he remained the same old Einstein, and his politics and sardonic nature made him a target for right-wing hatred. The world was listening to Einstein, and that made him a threat. The reason he was taken seriously was because of relativity. If his theory could be undermined, it would drain Einstein's political power. The theory of relativity was constantly under attack from physicists. This was not unusual; indeed, it was to be expected, as that is how science works. But now, with Einstein in the public eye, relativity also became a political target.

One of the early opponents of relativity was Einstein's fellow Berlin physicist Ernst Gehrcke. A serious scientist, he was convinced from the start that Einstein was wrong and began his campaign against the theory in the usual manner, publishing articles in technical journals reporting the flaws he claimed to have found. But these arguments were refuted one after the other by figures as prominent as Planck and Max Born. Sensing that the academic route was closed to him, Gehrcke moved from attacking Einstein in the professional realm to doing it in public.

Along with Paul Weyland, an antisemitic engineer with a desire for the spotlight, they formed the Party of German Scientists for the Preservation of Pure Science. They began with provocative charges leveled at Einstein in the popular press, accusing Einstein of academic misdeeds ranging from misrepresenting results to plagiarism.[4] Stirring up public sentiment, they arranged for a large event on August 24, 1920, featuring speeches from Weyland and Gehrcke about Einstein's flawed theory of relativity that were delivered to a packed house in Berlin's spacious philharmonic hall. The gathering was widely advertised and well attended, the audience largely composed of those sympathetic to the political leanings of its hosts, with antisemitic smear sheets circulated in the lobby before the talk. The audience also included a few physicists from the university, including Einstein himself, who was curious to hear what might be said.

Weyland contended that relativity was nonsense, "scientific Dadaism" that violated the German spirit and was accepted only because of groupthink in the scientific community. Max von Laue, who accompanied Einstein that evening, referred to Weyland as "the equal of the most unconscionable demagogue."[5] Gehrcke came up with no new objections but presented his old discredited views, explaining them for a lay audience without mentioning the errors that had been exposed in the journals. Einstein himself quite enjoyed the spectacle, laughing out loud, mocking the speakers, and clapping his hands along with the crowd.[6]

The circus was well received by those who were inclined to want Einstein's theory undermined but was roundly criticized by the local scientists who deplored the corruption of scientific discourse. Von Laue, Nernst, and Heinrich Rubens authored a widely read retort which not only addressed the problems with the physical arguments, but deplored the ad hominem nature of the attacks on Einstein.

Even Einstein himself felt the need to respond. In the *Berliner Tageblatt*, he published "My Reply to the Anti-Relativity Theory Company, Ltd," a condescending and sarcastic rebuttal in which he began by saying that neither of the speakers were in fact worthy of a response, but that he would give one anyway. He called attention to the flaws in their arguments and pointed out that their motivation was not scientific but political. "I have good reason to believe that there are other motives behind this undertaking than the search for truth. (Were I a German nationalist, whether bearing a swastika or not, rather than a Jew of international bent . . .)"[7] Einstein grouped together the attacks on relativity theory that he believed were not part of legitimate scientific conversation and included not only Weyland and Gehrcke but, surprisingly, Philipp Lenard as well. Recall that in the years leading up to and immediately after the development of his special theory of relativity in 1905, Einstein wrote laudatory things about Lenard and was extremely deferential in correspondence with him. But in this piece Einstein was dismissive: "I admire Lenard as a master of experimental physics; however, he has yet to accomplish anything in theoretical physics, and his objections to the general theory of relativity are so superficial that I had not deemed it necessary until now to reply to them in detail."[8]

After addressing all of these arguments, Einstein ended with a challenge: "Finally, I would like to note that, on my initiative, arrangements are being made for discussions to be held on relativity theory at the scientific conference in Nauheim. Anyone willing to confront a professional forum can present his objections there."[9] The conference was the annual meeting of the Society of German Scientists and Physicians, scheduled for September of 1920 in the central German spa town of Bad Nauheim. It was a serious gathering of important scientific minds, and Einstein had arranged for a session on the theory of relativity to be chaired by Planck. Eddington's results being newly published, Einstein's gravitational theory was a source of serious debate, and Einstein thought the time and location perfect to confront opponents in a professional setting. It was to be a high-level discussion for working physicists.

The buffoons from the philharmonic hall had cheated; they played without an opponent and without obeying the rules of scientific discourse. But real science is a tough sport, and you need to be well trained to set foot on the field with professionals. Set these clowns up against real scientists and they would either fold or get pummeled. Either way, such humiliation would finish them off for good. Not that Einstein thought he would actually see it. Most likely, they would pass on the offer. They knew they couldn't win the game fairly—that is why Gehrcke went from the journals to publicity stunts—and not taking up the challenge would show them to be the intellectual lightweights they were. Whether they showed up or not, the annoyance would be gone.

But that is not what happened. Einstein's challenge was taken up enthusiastically, and the antisemitic, pronationalistic opponents of relativity descended on Nauheim and crashed the meeting en masse. Local authorities were worried about violence breaking out around the session—something rarely observed at physics conferences. The angry crowd packed the hall, and jeers and chants disturbed the exchange between Lenard and Einstein. Felix Ehrenhaft, an Austrian physicist who was present at the scene, reported that Einstein was "interrupted repeatedly by exclamations and uproar. It was obviously an organized interruption. Planck understood this and was pale as death as he raised his voice and told those making the row to be quiet."[10]

Order was restored, and the session ended with some actual scientific discussion, but in the end both speakers left with changed perspectives. Lenard, already irate with Einstein for having lumped him in with Gehrcke and Weyland in his editorial in the *Berliner Tageblatt*, was thought by the other physicists to be the official mouthpiece of those who disrupted the meeting and was therefore dismissed as unserious himself. Lenard was being treated as less than a real scientist by his fellow physicists. For a Nobel laureate, this was an insult of the highest order. Lenard thought Einstein was wrong for legitimate scientific reasons, and as a result he was being ridiculed by his colleagues and branded a crank despite his lifetime's work, work that should have granted him a place of the highest respect. Einstein, relieved that the session had been saved to some degree, realized that the political objections to his work were not going away.

Indeed, the mass bullying that Einstein witnessed would become commonplace in Berlin in the coming years. As the economy went through its postwar contraction, antisemitism became rampant. A wave of eastern European Jews fled the pogroms in Russia and Poland. The combination of foreign-looking Jews in need of manual labor jobs at the bottom of the socioeconomic ladder, and wealthy Jews who were influential at the highest end of the corporate hierarchy, led to antisemitic conspiracies. If Germans were unemployed, it was because these Jewish refugees were taking the jobs of honest, hard-working Germans, jobs given to them by powerful assimilated Jews in industry and the Weimar government. Laws began to appear in localities so that *Ostjuden* (that is, eastern European Jews) were imprisoned "almost without cause."[11] Plans were created for mass deportation. Antisemitic incidents became common.

The economic difficulties of most Germans turned to complete loss, as the Deutschmark went into free fall in June of 1921. Hyperinflation over the next three years wiped out the wealth of all Germans except those who had their money in foreign holdings, generally those with international business ties. Again, the eyes turned to Jews, who were portrayed as bankrupting the nation for their own enrichment.

The plight of the eastern European Jews and the rising tide of antisemitism around him affected Einstein. He had renounced his religion when he first left Germany and repeatedly reasserted his alienation in official situations, but the treatment of the Ostjuden in both Berlin and abroad, especially in Ukraine, which he repeatedly labeled as "hell" in both public talks and personal correspondence, struck a chord. These were his "tribal kin" who were being so brutally repressed, and it awakened in him a sense of connection.

Observing the maltreatment of eastern European Jews at the university, Einstein opened his lectures to them, allowing them to audit his class without payment. This resulted in a protest led by antisemitic German students, who complained that their seats were being given to Jewish foreigners. They raised such a fuss in the classroom that Einstein was unable to teach. He was forced to discontinue the practice but instead held a special independent lecture series just for the eastern European Jewish students.

The situation led him to reassess his Jewishness, an awakening that left him open for an appeal from those championing the Zionist cause. Zionists, particularly Kurt Blumenfeld, the leader of the main Zionist organization in Berlin, were keen to reach out to Einstein because his name would be useful to the cause. But Zionism was not a natural fit for Einstein, who, to the core of his being, opposed every form of nationalism. Nation-states were artificial entities, and the patriotism they inspired was a closed-mindedness that served only to divide people for no reason, leading to nothing but discrimination and war. Nation-states were problematic, but it could not be denied, Einstein thought, that Jews across the world formed a nation in another sense. He had seen Ostjuden before, when he lived in Prague, and they seemed strange to him, living unnecessarily anachronistic lives. To Einstein they were a "them" and not an "us." But their plight led him to consider the peril of antisemitism, and he realized that it was not limited to the east in Russia, Ukraine, or Poland, but was also present in France and Germany in the west and seemingly everywhere else he considered. The experiences of Jews everywhere had core commonalities that united them into a nation.

Nations are artificial, Einstein argued, but the effect of being a part of a nation based on one's predilections, beliefs, and general stance toward the world is very much real. Jews had certain characteristics that came from their common experiences. Being a Jew is not part of the natural world—that is, it is not biological or intrinsic—but it is real. Cultures can create aspects of the personalities of the people by virtue of their organization and the distribution of social power. Oppressors and the oppressed will think differently. Jews in different parts of the world had different particular experiences, but among them existed a common bond that stemmed from the place that Jews everywhere have been forced to occupy. In this sense, Jews form an artificially constructed but very real nation spread throughout the world, and wherever they were, Jews were not safe.

In Germany, this lack of security led to very different approaches within the community. One stance was assimilationist —they hate us because we are different, so we will be safe if we become like them. Some of Einstein's friends were what we can call "strong assimilationists"; that is, they tried to become "purely" German by eliminating their Jewishness altogether. The chemist Fritz Haber came from a Chassidic family but converted to Lutheranism and did whatever he could to be as patriotically German as possible.[12] Haber worked at the highest levels of government open to a scientist, and during World War I led the German program that developed chemical weapons. Haber loved Germany and did everything in his power to receive its love as well. But Einstein disapproved, asserting that no matter how much he or any other strong assimilationists strove to deny who they were and adopt a new mode of being, they would never be accepted. Try as they might, the Germans they so longed to be one with would always exclude them.

A second route is "weak assimilation—that is, try to be authentically German *and* Jewish. Many of these groups shared the non-Jewish German disdain for the Ostjuden and strove to show that they as German Jews should be associated with the Fatherland and not with their eastern cousins. When one such group, German Citizens of Jewish Faith, wrote to Einstein in April of 1920 requesting his participation in a panel discussion of antisemitism, he declined, writing, "More dignity and independence in our own ranks! Not until we dare to see ourselves as a nation, not until we respect ourselves can we gain the respect of others, it must start with us then it will follow. . . . Can the 'Aryans' have respect for such sycophants?"[13] Assimilation, strong or weak, was a dead end, Einstein believed. Jews would not gain anything by rejecting the community in part or whole. German Jews would never be accepted by the non-Jews as "authentic." The self-loathing that was the result of the antisemitism must be rejected, not embraced. Its internalization could not be turned into anything healthy.

Jews must see all other Jews as their cousins, but one could not, as the German Citizens of Jewish Faith were trying to do, reduce Judaism to a matter of faith. "What are the characteristics of the Jewish group? What, in the first place, is a Jew? There are no quick answers to this question. The most obvious answer would be the following: A Jew is a person professing the Jewish faith. The superficial character of this answer is easily recognized by means of a simple parallel. Let us ask the question: What is a snail? An answer similar in kind to the one given above might be: A snail is an animal inhabiting a snail shell. This answer is not altogether incorrect; nor, to be sure, is it exhaustive; for the snail shell happens to be but one of the material products of the snail. Similarly, the Jewish faith is but one characteristic product of the Jewish community. It is, furthermore, known that a snail can shed its shell without ceasing to be a snail. The Jew who abandons his faith (in the formal sense of the word) is in a similar position. He remains a Jew."[14]

By doing their best to include themselves with the Germans to the exclusion of the Ostjuden, the German Citizens of Jewish Faith had inadvertently excluded Einstein. He had no "Jewish faith," so he was being asked to address the exclusion of Jews by a group that excluded him. The problem was not their definition, it was their impulse to exclude. Assimilationism in all of its forms, Einstein thought, was not only doomed to be ineffective, but dangerous to the Jewish community itself.

But Zionism also seemed to be complicated. For Einstein,

being a nation did not mean having a nation-state. The Zionist aspiration for a Jewish country in Palestine was one he never shared, and he worried that it would rob Judaism of its moral core. "The bond that has united the Jews for thousands of years and that unites them today is, above all, the democratic ideal of social justice, coupled with the ideal of mutual aid and tolerance among all men."[15] If Zionism became a movement that was focused on the idolatry of a particular piece of land, then the emergence of all of the evils that have plagued Jews across the globe for thousands of years would find a new source in Jews themselves. Turning Palestine into a Jewish state threatened the heart of Judaism. "My awareness of the essential nature of Judaism resists the idea of a Jewish state with borders, an army, and a measure of temporal power."[16]

Yet Einstein considered himself a Zionist. For him, the movement meant the development of a safe area in Palestine for Jews to evolve a cultural and intellectual base from which a progressive change could emerge in Jewish consciousness worldwide. Einstein's Zionism stemmed from three factors.

First is concern for the safety and well-being of the Ostjuden and all persecuted Jews around the world. When antisemitic violence occurs, Jews need a safe place to flee. This would be established in Palestine, not as an area taken from the non-Jewish Palestinians, but by living peacefully with them.

Second, Einstein thought that the establishment of a Jewish center of learning and culture in Palestine would serve the assimilated European Jews who were in danger not of losing their lives, but rather of losing themselves. In Europe, Einstein saw the more prosperous Jews suffering from a spiritual sickness, a sociological version of the Stockholm syndrome in which the Jewish people had found themselves to be denigrated so long that they internalized the antisemitic sentiments and sympathized with their persecutors. "When I came to Germany fifteen years ago I discovered for the first time that I was a Jew, and I owe this discovery more to Gentiles than Jews . . . I saw worthy Jews basely caricatured, and the sight made my heart bleed. I saw how schools, comic papers, and innumerable other forces of the Gentile majority undermined the confidence of even the best of my fellow-Jews, and felt that this could not be allowed to continue."[17] Jews believed themselves to be inferior in the way that the antisemites portrayed them, and they convinced themselves that, to be whole, they must lose their Jewishness, they must become like their tormentors. This could take the form of assimilation, and Einstein contended that antisemitism was the only force that has maintained German Jewry, who would have, but for the roadblocks of the antisemites, lost themselves completely.[18]

Third, Einstein came to realize that his own personal intellectual relationships with his Jewish friends were of a different type than those he had with his non-Jewish associates. He seemed to engage in different ways with those who had a Jewish background. He felt more open, more at home with these colleagues, despite his rejection of the culture for so long.

If this feeling could be generalized, and all Jews could have the opportunity to think alongside fellow Jews, the result could be liberating and exhilarating to the individuals and productive to the group and humankind at large. A Jewish center of learning and culture in Palestine, focused especially on a university and medical school, would allow Jews to do what they do best—discover and innovate. "The second characteristic of Jewish tradition is the high regard in which it holds every form of intellectual aspiration and spiritual effort. I am convinced that this great respect for intellectual striving is solely responsible for the contributions that the Jews have made toward the progress of knowledge, in the broadest sense of the term. In view of their relatively small number and the considerable external obstacles constantly placed in their way on all sides, the extent of those contributions deserves the admiration of all sincere men. I am convinced that this is not due to any special wealth of endowment, but to the fact that the esteem in which intellectual accomplishment is held among the Jews creates an atmosphere particularly favorable to the development of any talents that may exist."[19]

Jewish culture fosters an environment that gives rise to intellectual progress. If such an outpost could be developed, the advances in every art and science would be notable, and Jews across the globe would see themselves in a different light. Bringing Jews together to be Jews together—as Einstein understood Judaism—could have only positive results for those Jews who participated, for the Jews who witnessed from afar, and for humanity as a whole. Not only would there be great developments that would save lives and better human existence, but a proud, self-possessed Jewish population who contributed to the betterment of all would undercut the internal and external sources of antisemitism. Jews would been seen and, more important, see themselves, as valuable.

This was not the vision of all Zionists, but they welcomed Einstein's proclamation of association. Einstein's status as a celebrity and his embrace of Zionism could only help the cause—if managed properly. Blumenfeld and Chaim Weizmann both realized that Einstein could be a bonanza for the Zionist movement, in terms of both publicity and fund-raising. They were under no illusions concerning Einstein's unorthodox brand of Zionism. Indeed, in one letter to Weizmann, Blumenfeld explicitly stated, "Einstein, as you know, is no Zionist,"[20] but that would not matter. They were not looking for Einstein's leadership or ideas. They wanted his name, not his brain.

Einstein's Zionist vision centered on the establishment of the Hebrew University in Jerusalem, and Weizmann and Blumenfeld knew to keep him focused on that. Doing so would not only keep Einstein from making statements that conflicted with the political views and plans of the Zionist movement, but could be very useful in raising funds from non-Zionist Jews who would never give to Weizmann's organization, but likely would contribute to an academic endeavor of which Einstein was the public face. Einstein's imprimatur would give the project prestige, necessary for a not fully existent university that did not yet have a full faculty to brag about. Einstein would serve as a proxy for the scholars who would eventually work there. There were hopes that Einstein himself would join that faculty, and hints to that effect were regularly dropped before potential donors, but Einstein himself never seriously considered it because he regarded himself as essentially European.

Einstein's enthusiasm for the Hebrew University project led to his first major Zionist work, a fund-raising trip with Weizmann to America in the spring of 1921. The trip was somewhat complicated by the fact that Einstein had received several offers to speak in the United States just before agreeing to accompany Weizmann. On the advice of friends, he had responded to the American universities who inquired—Princeton University and the University of Wisconsin—that his speaking fee would be $15,000. At the time, this was an exorbitant figure. Einstein requested it for two reasons. First, he did not want to go to America. The trip would be too much for him given his health and everything else on his plate. But he did have his children to consider, and if the American universities were willing to pay such an absurd amount, he would do it for the money. Both universities declined, expressing shock that Einstein would request such an amount. When Einstein agreed to accompany Weizmann, he decided that he would intersperse his Zionist addresses with university lectures to raise money for his family at the same time. But the presidents of Princeton and Wisconsin put the word out that Einstein was greedy, arrogant, and self-serving and that they would do well to avoid inviting him. Einstein acquired a reputation in line with the standard stereotype of Jews.

It was not the first time. When he published his article in 1920 taking Gehrcke and Weyland to task, a number of his colleagues objected. They were sympathetic given the absurdity of the charges against him, but they saw Einstein taking the spotlight in a way that seemed inappropriate for a working scientist. Their job was to discover universal truths, not find personal glory. The coverage that Einstein received and now was actively seeking came across as unseemly to them, inappropriate for someone of his place and occupation.

This chorus became louder that year when Alexander Moszokowski, a journalist and acquaintance of Einstein's, decided to write the first biography of the scientist and title it *Einstein, the Searcher.* He claimed correctly that it was based on conversations with Einstein, who had no problem with the book and no qualms about helping a friend. But other academics were scandalized and warned him that if he did not stop the publication or at least repudiate the book, he would be playing into the hands of the antisemites who were seeking to label him as a self-aggrandizing Jew. Max Born and his wife went so far as to urge him to get a restraining order and make sure that fact was reported in local papers. If not, they contended, "your Jewish 'friends' will have achieved what a pack of anti-Semites have failed to do."[21] Alongside the "Einstein, modernist Saint" narrative, the "Einstein, self-promoting Jew interested only in commercial gain" story developed as Einstein became part of the collective consciousness.

Einstein was able to smooth over relations with the American universities, and although it required a schedule that came together uncomfortably late, the trip proceeded. When their ship arrived in New York on April 2, 1921, a tremendous crowd led by a slew of reporters greeted Weizmann, Einstein, and Elsa. Einstein, with his rumpled dress and wild hair and his quick-witted, offhand responses to questions, was exactly what the press had hoped he would be. Einstein's quips were eagerly collected, although the best line of the day came from Weizmann, who when asked if he now grasped the content of the theory of relativity, replied, "During our crossing, Einstein explained his theory to me every day, and on our arrival I realized that he really understands it."[22]

But what Einstein did not understand were the internal Zionist politics behind their trip. Their tour was not as simple as Weizmann and Blumenfeld had led him to believe. It was not just a fund-raising affair, but part of a larger turf battle, an attempt by Weizmann to secure his place as the undisputed leader of the worldwide Zionist movement. In Europe, he called the shots; but in America, there was another group, led by Louis Brandeis, and it resented Weizmann trying to run roughshod over their efforts. The group saw Weizmann's use of Einstein as a publicity stunt to draw attention away from his real aim—taking pro-Zionist donors for his own, thereby drying up Brandeis's cash flow and hence subjugating the American Zionists to the will and ideas of the Europeans.[23] Brandeis's group did not idly stand by, but rather tried to stop the defection of donors by publicly accusing Weizmann's people of misappropriating funds. Weizmann's trip was provocative, and the situation was tense.

In this way, Weizmann worried about Einstein on two grounds. First, he knew that Einstein was a loose cannon, that his views on Palestine differed from those Weizmann was trying to push and that Einstein could say things that could embarrass the cause. Second, he wanted to make sure that Einstein remained unaware of the internal Zionist conflict. So Einstein was asked to keep his comments at the events brief and general. Einstein obeyed and regularly referred to Weizmann as "our leader," and deferred to him on questions of Zionist matters in his public comments. Further, Weizmann made sure that the Einsteins—he worried about Elsa as well—had minders, official companions from the European delegation wherever they went who had control over what they saw and with whom they spoke.

This plan lapsed when they briefly split up, Weizmann heading off in a different direction to take care of some business while Einstein traveled to Washington, D.C., to speak at the National Academy of Science. With Einstein unattended in Brandeis's backyard, the Supreme Court justice found time to meet with Einstein and fill him in on the split from his perspective. Einstein played diplomat, sympathizing with Brandeis in Washington but claiming to be on Weizmann's side when speaking with him, Blumenfeld, or any other member of the European contingent. Einstein thought himself a mere "prized ox" to be paraded around by the Zionists and was happy to be such as he advanced his own interest in the Hebrew University while remaining above the infighting.

While the trip failed to raise anywhere near the amount of money that Weizmann and Blumenfeld had hoped, an unintended result was to strengthen Einstein's identity as a Jew. "It was in America that I first discovered the Jewish people. I have seen any number of Jews, but the Jewish people I had never met either in Berlin or elsewhere in Germany. This Jewish people, which I found in America, came from Russia, Poland, and Eastern Europe generally. These men and women still retain a healthy national feeling; it has not yet been destroyed by the process of atomisation and dispersion. I found these people extraordinarily ready for self-sacrifice and practically creative."[24]

On the way back to Germany, Einstein spent time in England giving talks at universities. His reception was quite different from the one he received in America. The scars of the war were still fresh, and British conservatives were not about to let any German speak in England, much less the man who displaced Newton. To try and counter this attitude, his hosts made clear that while Einstein lived and worked in Berlin, he was, in fact, Swiss, and anyway held positions on the war that should separate him from other Germans. But it was Einstein himself who persuaded his audiences. When introduced in London, neither he nor his host received even polite applause; instead they faced an uncomfortable stony silence. Einstein, speaking in German and using a translator, began with kind words for Newton and proceeded to be Einstein, speaking without notes and without pause, being passionate and clever. After an hour, he apologized for the length of his comments and was greeted with a rousing standing ovation in the hall.[25] Relativity as diplomacy was an effective tool for international peace in Einstein's capable hands.

But if the trip left Einstein feeling good about his relation to Jews and his role in Germany's reintroduction into the geopolitical dialogue, at home the nationalists had a very different view. Fritz Haber had warned Einstein that his sailing on a Dutch ship to America and England just as President Harding had stepped away from moving the Versailles accord through the U.S. Senate, and as the Brits were increasing the severity of sanctions against Germany, was being painted as evidence of his disloyalty.[26] Einstein was not only hobnobbing with the enemy, but doing it to raise funds for Jewish projects in Palestine. He was raising large sums of money from the countries that were bankrupting Germany not for the purpose of easing German suffering, but for helping Jews leave Germany. This picture fit, Haber contended, with the nationalist contention that Jews are not loyal patriots but have concern only for their own well-being and care nothing for the Fatherland.

These concerns were on Einstein's mind when he received an invitation in 1922 from Paul Langevin and Marie Curie to speak in France. While the United States and Great Britain were also on the other side in the war, France held a particular place of derision in the German consciousness. If Einstein was being given a hard time in the press for his trip with Weizmann, going to France would cause hysterics. He politely declined at first, citing the political situation. But then he spoke with his friend Walther Rathenau.

In certain ways, Rathenau was not the sort of person with whom Einstein generally preferred to spend time. Born extremely wealthy, the son of the founder of Germany's General Electric, Rathenau eventually took over the major corporation. As a captain of industry, he wielded significant political power. He bragged that three hundred men controlled the direction of Europe, clearly implying that he was among this elite group of plutocrats. The claim was not unusual coming from Rathenau, whose personality was strong to the point of conceit. Like Haber, he was a Jew whose deepest desire was to be accepted as authentically German. Rathenau was a patriot and did whatever he could to advance the interests of the Fatherland at peace and at war.

But Rathenau had another side. He was not a mere capitalist with political ambition, he was a renaissance man: cultural critic, philosopher, and writer. He was a central player in the Berlin avant-garde scene, and his portrait was painted by Edvard Munch. Like Einstein, he had a broad mind and thought himself a public intellectual.

The two met at a dinner party, and Einstein was so taken with him that he invited Rathenau to his home for an evening meal and conversation. It was an opportunity Rathenau eagerly accepted, and they became friends. Einstein thought that Rathenau had "an elegant and sparkling spirit."[27] In the other direction, "it was a unique relationship, perhaps the only one in which the arrogant and exceptionally successful Rathenau acknowledged someone else's intellectual superiority," wrote Einstein biographer Thomas Levenson.[28]

Rathenau had been one of the leaders in the formation of the new Democratic Party, which Einstein wholeheartedly supported. Indeed, Rathenau's involvement (he gave both his time and his money) was one of the reasons German conservatives labeled it as "the Jewish party." Rathenau had a minor role in the government helping to oversee postwar reconstruction. He knew that a solution to Germany's problems required a renegotiation of the reparations agreed to at Versailles. He also knew that France was a major hurdle to accomplishing this. So when he heard of the possibility of Einstein traveling to Paris for public talks, he knew that this was a good opportunity for Germany. Distrust on both sides ran high, and Einstein would be a wonderful cultural diplomat. Anything that could increase goodwill could only be helpful, and Rathenau told Einstein that it was his duty to make the trip and give the talks. His words were effective, and Einstein contacted Langevin and Curie and told them of his change in heart, but he added that he would not bring his wife so that the trip would look more professional and less political.

Einstein visited France in March of 1922. It was not as successful as his appearances in Britain, but it gave him a chance to meet with not only his colleagues, but his old friend Solovine from the Olympia Academy from his Bern days. On the way back to Germany, at his request he, Solovine, and Langevin were taken on a deeply moving tour of battlefields of the Great War, still marring the countryside.

When he returned home, he contacted Rathenau to discuss the trip. The meeting was urged by Blumenfeld, who had an ulterior motive. He wanted to use Einstein's friendship with Rathenau to gain a private audience and convince Rathenau to step down from the post of foreign minister that he had recently accepted, the first Jew to ever hold so high a post in the German government. On the one hand, it was remarkable that a Jew could even be considered for the role, much less installed. But from the Zionists' perspective it was not a sign of progress; rather, it was bad for the Jews to have such a flamboyant character in so public and powerful a place. Rathenau was not just any Jew; he was a wealthy industrialist who insisted on occupying the public spotlight. As foreign minister, his job would be to work closely with representatives of France, Britain, and the Soviet Union. Indeed, Rathenau himself thought that the best position from which to renegotiate Germany's position would be by establishing warmer relations with the Soviets. The split between France, Britain, and America on one side and the USSR on the other was a possibly fortuitous situation that Rathenau thought he might exploit to Germany's advantage. But this fed into every antisemitic stereotype the nationalists were employing—that the Jews were rich, power-hungry opportunists with Communist sympathies who were more concerned with their own enrichment than with the country's well-being. In Blumenfeld's mind, Rathenau had to step down. Einstein did not share Blumenfeld's position, but he did worry about Rathenau's personal safety. It was a concern that Rathenau would have done well to consider.

The depreciation of the currency and the accompanying hyperinflation were wreaking havoc on the German economy. To these financial problems were added an additional insult to national sentiment. To end the war, the Germans had agreed to disband most of their army but for small groups needed to maintain internal order. Foreign troops, largely French and Belgian, occupied large areas of German soil, while Germany was left unable to defend itself. This rankled nationalists as a whole, but it especially angered the troops who returned home from the war. Their comrades had died and they had endured unspeakable conditions in the trenches to keep the enemy off of their own soil, and here was a "Jewish government" agreeing to allow enemy troops to occupy the Fatherland. It was infuriating.

From these returning troops, many of whom were unable to find a job, *Freikorps* units—paramilitary cells—were created. They quickly became a place of refuge for those with a strongly nationalistic ideology and a seething fury at the status quo. They would serve the centrist government when it meant putting down left-wing uprisings such as that led by the Spartacus League, which sought to replicate the Bolshevik revolution in Russia, but that did not mean they were loyal to the government. Indeed, since they were informal paramilitary groups, many members felt that their duty was to the Fatherland (as they understood it), not to the country as led by the Weimar government.

On the morning of June 24, 1922, four members of the Freikorps pulled their car alongside Rathenau's open convertible as he was on his way to work. With a machine gun and a hand grenade, they assassinated Germany's foreign minister. Rathenau's murder signaled a major change; the culture struggle was no longer metaphorical. Einstein recognized this and knew which side he was on. He began receiving death threats himself, warning him that what happened to Rathenau could be awaiting him as well.

Einstein considered leaving Berlin and spent time away from the city, relocating temporarily to Kiel, but ultimately he decided to stay and instead relied on a misinformation campaign stating that he had fled, while he actually just kept a low profile. He canceled most public appearances and resigned his post on the League of Nations' International Committee on Intellectual Cooperation. This period was extremely stressful for the Einsteins, who felt concern for their safety every time they stepped out.

Ultimately, they would soon be leaving anyway on an extended tour of Japan. This was when word of his Nobel Prize was imminent. Instead of remaining to receive the award, he chose, for reasons of personal safety and because the ceremony for the accolade was of little interest, to leave for Asia in October of 1922. Einstein greatly enjoyed his trip. The six-week journey took them through the Suez Canal to Sri Lanka, Singapore, Hong Kong, and Shanghai before arriving in Kobe. He loved Japan, and Japan loved him. He marveled at the culture and was received like a hero everywhere he went. The contract he had signed was with a company that considered the trip purely a matter of public relations for their own marketing purposes, and the small print barred him from delivering lectures the company did not arrange. But the six-month respite from the turmoil in Germany suited Einstein just fine.

Because the return trip would take him back through the Suez Canal to reach Europe, Einstein thought it would be a good time to arrange his first visit to Palestine. It would be only a week's stopover, far too short in his mind to be a real visit, but given all of the Zionist work he had been doing it seemed inappropriate for him not to at least briefly see the land and talk with some of the people. Weizmann and Blumenfeld were more than happy to make the arrangements, and in 1923 Einstein arrived in Palestine.

Both Albert and Elsa were again to be accompanied at all times by officially approved guides during their entire stay in Palestine so that they would be shown and told exactly what the Zionists wanted them to see and hear.[29] Einstein would give a speech ceremonially launching the Hebrew University, but he would also be told exactly what the official stance was on various issues. Again, the Zionist leaders saw the opportunity as propagandistically powerful, but also worried about Einstein saying what they did not want said.

Arriving in Jerusalem, Einstein was the constant guest of Zionist and British officials at receptions big and small. He walked the old city and took in the sites. He visited kibbutzim and toured Tel Aviv. While he found the former to be impressive, it was the city that affected Einstein the most. Agriculture was one thing, the Arabs could do that just fine, but Tel Aviv was a wonder. Here was a city of Jews, a contemporary urban environment created out of nothing but vision and sweat. It was the modern world rising out of desert, brought into being purely through everything that Einstein thought to be the essence of Judaism for the future—science, art, community, and creativity. Tel Aviv represented exactly the picture he had envisioned of what the Jewish presence in Palestine could be. "The accomplishments by the Jews in but a few years in this city elicit the highest admiration. A modern Hebrew city with busy economic and intellectual life shoots up from the bare ground. What an incredibly lively people our Jews are!"[30]

It contrasted with his feelings for the ultraorthodox he saw praying at the Western Wall, whom he called "dull-minded tribal companions . . . men with a past but without a future."[31] If Zionism was to succeed in curing the sickness of the Jewish soul, it must embrace Judaism moving forward, Einstein thought, not looking backward.

Einstein's lectures in Jerusalem were major events. He spoke on Mount Scopus, where the Hebrew University would be erected. British, Jewish, and Arab dignitaries of the highest ranks were invited, although the Arab delegation declined the invitation. He had a sentence translated into Hebrew with which to begin his comments: "I, too, am happy to read my address in the country whence the Torah and its light emanated to all the enlightened world, and in the house, which is ready to become a center of wisdom and science for all the peoples of the east."[32]

He gave the rest of the talk in French, explaining that he regretted his inability to speak the language of the university.[33] He spoke for an hour and a half and received a rousing ovation. The next evening he spoke again, this time in German—it was the first time since the First World War that a public talk was given in Palestine in the German language.[34]

Einstein left enthused about the Zionist project. To Solovine he wrote, "I greatly liked my tribal companions in Palestine, as farmers, as workers, and as citizens. The land, on the whole, is not very fertile. It will become a moral center, but it will not be able to absorb a major part of the Jewish people. On the other hand, I am convinced that colonization will succeed."[35] Einstein's commitment to Zionism grew.

Einstein's identification as a Jew, however, was intensified elsewhere as well. The mid-1920s saw an end to the hyperinflation, and while the economic and political situation seemed more under control, beneath the surface was a simmering tension. The nationalists were gathering, and their rhetoric was intensifying. Brown-shirted thugs were causing disturbances as a formal political structure for their anger was being created by Adolf Hitler and his early followers. Combining promilitaristic ideas with expressed antisemitism seen through a romanticized mythology of German identity, National Socialism grew slowly, seeming more a nuisance than a true threat to democratic ideals.

But if Einstein had already been on their enemies list, he became even more of a threat as his fame grew and his political statements became more pointed. Einstein continued to travel the world, making trips to North and South America, and wherever he went he continued to be trotted out as a national treasure by local German dignitaries despite the fact that many of them adhered to the increasingly nationalistic viewpoint he so despised. Amused, Einstein likened himself to a foul-smelling flower that the Germans insisted on repeatedly putting in their buttonhole.

As the climate grew worse, the attacks on Einstein started to come from the place he least expected—the scientific community itself and in particular two of his former allies. As the Nazi worldview began to creep in from the margin of the political spectrum, everything became viewed through a racist lens. Einstein's theory of relativity was termed "Jewish science" by those championing National Socialism. To be fair, all of modernist culture was painted with a broad brush as contrary to true

German values, but Einstein's work received special treatment from Philipp Lenard and Johannes Stark, who both had been enthusiastic supporters of Einstein decades earlier. Now they were the leaders of the "Aryan physics" movement.[36] "'German physics?' you will ask.—I could also have said Aryan physics or physics of the Nordic type of peoples, physics of the probers of reality, of truth seekers, the physics of those who have founded scientific research.—'Science is international and will always remain so!' you will want to protest. But this is inevitably based upon a fallacy. In reality, as with everything that man creates, science is determined by race and blood . . . Nations of different racial mixes practice science differently."[37]

For Lenard, the route leading to his adherence to the Aryan physics movement began with his public humiliation in Nauheim. He had always been a patriotic German, but not one particularly inclined toward antisemitism. His own dissertation director, Heinrich Hertz, was half-Jewish, and it was a point of deep pride in Lenard that he was among the great Hertz's intellectual offspring.

As did so many others, he had bought into the war frenzy during the First World War, and he purchased war bonds to support the Kaiser's noble fight. His entire savings went to the effort. But when the war went badly and the Kaiser was replaced with the Weimar government, the hyperinflation made the bonds worthless. He was a Nobel Prize winner, a world-renowned scientist, and now he was broke, his life savings wiped out because of the "Jewish government." To Lenard, the symbol of this corrupt administration was Rathenau, whom he hated. When word of the assassination was announced, Lenard refused to lower the flag at his institute in tribute. A group of leftist students, angry at the lack of respect, attacked Lenard physically, leaving him feeling helpless, vulnerable, and angry.

These negative feelings were exacerbated by resentment

that Lenard harbored from a professional dustup. His research project on cathode rays was also the subject of experimentation by the British physicist J. J. Thomson. While both he and Thomson were awarded the Nobel Prize, Thomson had become much more famous and well regarded in the scientific community for doing what Lenard saw as the same exact work. Indeed, Lenard claimed that he was the first to publish one of Thomson's most important results. Thomson had known about Lenard's paper and failed to cite it, an oversight that Lenard believed was not an accident, but rather academic misconduct so that Thomson could claim priority for the discovery. When Lenard wrote to him and politely pointed out the lack of attribution, Thomson added a footnote, but only a reference to a later result and left Thomson's priority unchallenged. Lenard was irate—not that he was interested in personal glory; as a good German, his interests were focused on the good of the whole not on the individual.

In Lenard's mind, to be self-aggrandizing and caring about utility to one's own self is an ineliminable part of the British character. Look at the writings of English thinkers in economics, political theory, and ethics; they are all about costs and benefits and how they help or hurt the individual. The British are self-serving and morally bankrupt, Lenard thought, and he loathed not only Thomson personally, but Great Britain generally. So much the better when they were the enemy in the war; he now had reason to hate them. After the war, there was Einstein, who not only went to London to be patted on the back, but was like them all the time. Einstein was not a proper German; he was selfish and sought the spotlight for himself. He was therefore British in affect, and the hatred Lenard already felt could be transferred to Einstein, hatred that came with a narrative that meshed well with those being peddled by the Nazis.

Stark's path was similar. With a more rightward-leaning

political orientation and his base at the University of Greifswald, a quite conservative setting, Stark always had a predilection for the nationalistic. But he initially had no antisemitic or anti-Einstein bias. Indeed, he was one of Einstein's original supporters after the 1905 paper, and it was Stark who invited Einstein to write a survey article on the contemporary literature and research projects concerning relativity in 1909, ultimately leading to the general theory.

But the relationship changed soon after. Independently, Stark and Einstein each discovered a relationship between light absorption and chemical reactions. We now call this the Stark-Einstein law, but in 1913 Stark was certain that he got there first and therefore should have priority for the result. Further, his derivation was more elegant and should be the canonical approach. When he saw that Einstein was getting credit in the scientific literature, or that both names were being mentioned, Stark complained, but Einstein was dismissive. His attitude conveyed that he was above petty issues such as priority, and Stark should be ashamed of himself for being concerned. But this was not how Stark saw it. It was a conspiracy. Einstein, the celebrity, could play the saint, keeping his hands clean because he had his cronies doing his dirty work, undermining the legitimate claims of others who bested Einstein scientifically. The game was rigged.

This view was reinforced when Stark applied for an opening later that year for a physicist at Göttingen. It was a much more prestigious setting than Greifswald, and Stark was convinced that the job was his for the asking. Academic advancement in Germany was an explicitly political matter. While universities and departments could make requests and suggestions, it was a central bureaucrat who made the appointment. In any field, the researchers could be ranked, and the top of the list would receive preferential treatment for appropriate openings. The Göttingen job should go to him.

But it did not. Instead, it went to Peter Debye, a physicist from Holland who never adopted German citizenship. Stark took the decision as a personal snub. Here was a position at one of Germany's premier universities, and they passed over a leading light of German physics for a foreigner. Debye's adviser had been Arnold Sommerfeld at Munich, a friend of Einstein's. The Einstein syndicate had struck again. Stark was beside himself and lashed out at the "Jewish and pro-Semitic circle" who controlled German physics, labeling Sommerfeld as their "enterprising business manager."[38] What the Jews were doing in the larger economic and governmental realms, Stark contended, they were also doing to academic science—seizing power for their own advancement and leaving the good, noble patriotic Germans to suffer for it.

Lenard and Stark combined to lead the Aryan physics movement, which was elevated in standing with Hitler's rise to power. Einstein's theory of relativity was to be rejected because it was "Jewish science." It had been accepted widely not because it had been justified on the basis of the evidence, but because a Jewish cabal had taken control of the scientific infrastructure and demanded it. Einstein was not a genius, but a scheming front man for this pernicious enterprise that was destroying science the way that Jewish-led movements were simultaneously destroying art, music, mathematics, architecture, and every other human endeavor. It necessitated a radical solution, cutting the rot out at the roots.

The rise of the right wing in German politics made life complicated for Einstein. The year 1929 saw his fiftieth birthday, and as a gift friends pitched in and bought Einstein a new sailboat. Sailing had become his favorite pastime. Wearing a kerchief knotted at the four corners as a hat, he would methodically work the sails deep in thought. He loved the sun and the wind, but most of all the solitude sailing afforded him. Einstein and Elsa had a small country shack beside the Havel

River, from which he would set out. The place was small and ramshackle. Elsa refused to spend more than a night or two in the rough surroundings, leading Einstein to call it "the castle."

Knowing his love for sailing, members of the municipal council in Berlin decided to give Einstein a new small house on the bank of the Havel.[39] Grateful for the gesture, Elsa wanted to see the new abode before accepting the gift, hoping it would be more to her liking. It was not. The council looked for a new location, and when nothing acceptable was found, the idea was floated to buy a piece of land and have an appropriate house built. Conservative members objected that Einstein should not be honored and that municipal funds certainly should not be used in any case. Eventually the matter was dropped and the Einsteins built their own country house on the lake in Caputh. The house became Einstein's fortress, the place to which he would escape when he needed to wall himself off from the chaos that Germany was descending into, especially after the crash of 1929.

But as Hitler's influence grew, Einstein began to sense that this hideaway was not enough. He had received an offer to spend the winter at the California Institute of Technology in Pasadena. The idea of being away without having to tour was attractive, and so he and Elsa wintered outside of Los Angeles.

One of the advantages of Cal Tech was that it was close to the Mount Wilson observatory, where some of the most important observational work in astronomy was being done. On a tour of the massive telescope and its facility, the guide explained to Elsa that this mammoth instrument was used to uncover the deepest mysteries of space, such as determining the shape of the universe itself. Unimpressed, Elsa replied, "Oh. My husband does that on the back of an old envelope."[40]

The Einsteins enjoyed their stay in Pasadena and were delighted when the university extended the offer for the next three winters. It was an attractive invitation. Good colleagues, a lack of responsibilities, and political calm in the United States would let him work while still maintaining his life in Berlin for the majority of the year. And so Einstein continued this pattern in 1931, 1932, and 1933. When nationalists tried to paint this as an abandonment of Germany, Einstein assured everyone that he still lived in Berlin, that he would be in Germany for the majority of each year, including the next year—except that that particular year happened to be 1933.

6

In Exile

Einstein was in California in January of 1933 when Hitler assumed power as the chancellor of Germany. Einstein had been in Germany for most of 1932 but had sequestered himself in Caputh, avoiding the degenerating street chaos of Berlin. When he and Elsa closed up the summer home to leave for their now-routine trek to Pasadena, Einstein told Elsa to take a good look, as they would never see it again. He was right. When Hitler took over, Nazis looted the house, purportedly to look for weapons and other evidence of wrongdoing.

Keeping track of events from abroad, Einstein made clear that he would never return to a Germany under Nazi control. He formally resigned from the Prussian Academy of Science. "The conditions at present prevailing in Germany induce me to lay down herewith my position in the Prussian Academy of Sciences. The Academy provided me with the opportunity to devote myself to scientific work, free from any professional obligations. I realize the great measure of gratitude I owe to it. It is with reluctance that I leave its circle, also because of the stimulation and the beautiful personal relations which, during that long period of my membership, I enjoyed and always greatly appreciated. However, dependence on the Prussian government, entailed by my position, is something that, under the present circumstances, I feel to be intolerable."[1] The Academy accepted his resignation with certain members upset that it arrived before they could remove him themselves. Ernst Heymann, the Academy secretary and a Nazi sympathizer, recorded the affair and added in the official record that Einstein was unfit for membership given his "atrocity propaganda" against the Reich and that the Academy had no regret concerning his resignation. Von Laue spoke up to have these comments removed, but no one—not Planck, Haber, or any of the others supported the motion; indeed, to the contrary, Heymann was praised for his "appropriate actions."[2]

Einstein was no longer welcome in Germany, even by the majority of his scientific colleagues. The feeling, of course, was mutual, and he decided to take action. As part of his typical itinerary when returning to Germany from America, he would disembark in Holland, where he usually took the train back to Berlin. This time he traveled by car to Brussels, where he went to the German consulate and surrendered his German passport and once again formally renounced his German citizenship. He wanted nothing to do with a Nazi-run Germany, a fact the conservative papers played up as they published articles brutally attacking Einstein.

In addition to ransacking his home in Caputh, the Nazi government seized his bank account and the thirty thousand marks it contained. Einstein was bothered but not excessively so; since the economically turbulent days of Weimar, he had made a habit of depositing all of his foreign earnings in banks outside of Germany, so he was well enough off despite the loss.

One of Hitler's first efforts was to enact the Law for the Restoration of the Professional Civil Service on April 7, 1933, removing Jews and political enemies of the Nazi Party from government jobs. Since the universities and the Kaiser Wilhelm Institutes were all under the federal umbrella, this meant that scientists were part of Hitler's first purge. The original order had been weakened to include an exception for those who fought at the front in World War I or who were the father or son of someone who died in battle—something deemed as proof of patriotism. The exception was demanded by Paul von Hindenburg, the German president who, while a thoroughgoing conservative nationalist and former general and war hero, loathed Hitler, and many hoped he would serve as a check on Hitler's power.

Under this exception, Haber was allowed to keep his position as head of the Kaiser Wilhelm Institute for Chemical Research. He was, according to the rules for implementation of the law, considered a Jew despite his conversion, but his efforts for the military were deemed sufficiently important to maintain his Reich citizenship. But having to cleanse his institute of all other Jews was too much, even for the enthusiastically patriotic Haber. A combination of his conscience and seeing the writing on the wall led to his resignation, and he left his beloved Germany to move to, of all places, Palestine. Haber, who as an anti-Zionist had quarreled with Einstein, urging him not to travel with Weizmann, and who had defended German motives and interests at every turn, was now offered by none other than Weizmann himself the directorship of what would become the Weizmann Institute of Science, an institution of higher learning dedicated to advanced scientific research. Haber would be the head administrator for a university in Palestine designed to foster the growth of Jewish education and science—exactly the idea Einstein worked to instantiate, and the one to which Haber so vehemently objected. The irony was not lost on Einstein, who wrote to his friend, "I am delighted to know that your former love for the blonde beast has somewhat cooled off. Who would have thought that my dear Haber would turn up here as the champion of the Jewish, and indeed the Palestinian cause."[3] Haber, however, never assumed the post, as he suffered a heart attack en route and passed away in January 1934 before leaving Europe.

For several months in 1933 Einstein stayed in the Belgian seaside town of Le Coq when he was not traveling to speak. He went to England for a couple of weeks to deliver the Herbert Spencer Lecture at Oxford, where he met with Abraham Flexner, who was representing Louis Bamberger, a Jewish philanthropist who had an idea for a scientific institute. Flexner was seeking out the most prestigious living scientists to begin envisioning what such a place might look like. The conversation with Einstein was not intended to recruit him, but both Flexner and Einstein were so excited by the possibility of what was shaping up that neither could deny the interest that each had in the other.

By now Cal Tech had a long-standing relationship with Einstein, who was pleased with the arrangement. Indeed, the school was hoping to secure Einstein's presence on a long-term basis. Einstein shared this desire as well. California suited him. During his weeklong stay in Manhattan before his ship turned south toward the Panama Canal on its way to California, he faced a constant frenzy of large crowds, including overexuberant and underintelligent reporters bombarding him with inane questions, and others requesting that he give talks before inevitably packed audiences. In Pasadena, on the other hand, the people were relaxed, and while he received many requests for autographs, they came with easy smiles.

When he ventured into Los Angeles proper, he would occasionally experience the mania he so disliked. One such instance occurred when the Einsteins accompanied Charlie Chaplin to the premier of his film *City Lights.* As they stepped out of their car they were mobbed by cheering, adoring fans. Einstein, stunned by the reaction, turned to Chaplin, who remarked, "They cheer me because they all understand me, and they cheer you because no one understands you." Einstein asked him, "What does it all mean?" To which Chaplin calmly replied, "Nothing." Such occasions, however, were the exception in California.

He maintained his acquaintance with Chaplin, who invited the Einsteins to his house for a private showing of *All's Quiet on the Western Front*, a film adaptation that had been banned in Germany. Chaplin's politics were very much in line with Einstein's, as were those of Upton Sinclair, muckraking journalist and author with whom Einstein also became friendly. At one gathering, Sinclair's wife challenged Einstein's theological views, as the Sinclairs both were believers in the occult. Parrying the attack, Elsa stepped in and said to the delight of all present that with respect to God, "My husband has the greatest mind in the world . . . but he doesn't know everything."[4]

While he had made friends with notable figures of the left, California remained, like the rest of America, out of line with Einstein's political leanings. Robert Millikan, the Nobel laureate and patriarch of the Cal Tech physics department, was extremely conservative and, while always respectful, clearly not happy with Einstein's political sentiments frequently published to great fanfare. Similar feelings were shared among the many wealthy potential donors to Cal Tech whom Einstein met at the many social occasions set up for him—the price he had to pay for the arrangement.

But this mingling which Einstein disliked would not be a part of the position that Flexner had in mind. Bamberger's Institute for Advanced Study would be located in New Jersey, in the town of Princeton, but unaffiliated with the Ivy League university there. It would be dedicated to purely theoretical science—no laboratories allowed. Like the museums of classical Alexandria, it would simply be a gathering place for the great scientific minds to meet and work at their own pace, with no expectations, allowing inspiration to strike when it would.[5]

Einstein was intrigued by the idea of being paid to be left alone to pursue his ideas. The negotiations with Cal Tech had recently turned rocky, the salary Einstein understood to be promised him verbally was greatly diminished when the written contract arrived. So when Flexner asked Einstein about his salary requirements for working at the Institute, Einstein lowballed the amount, asking for $3,000 a year. Observing that Flexner was taken aback by the request, he quickly offered to accept less for his modest living needs. He had misinterpreted Flexner's reaction; Flexner, shaking his head, made a counteroffer that included an annual salary of $10,000, coverage of Einstein's taxes, and an allowance for Elsa's travel costs. Einstein agreed and thanked him, although he would again offer to decrease the amounts before signing on the dotted line.

Princeton was not Pasadena. It was on the East Coast, where the weather was less hospitable and the people were less laid back, but the town offered other amenities. Between the Institute and the university, there would be ample intellectual engagement, and the community was home to a number of German, especially German-Jewish, refugees including his friend Paul Oppenheim, a chemist who had become quite wealthy as a part of the German chemical industry. Oppenheim used his money to acquire a magnificent collection of impressionist art, which was shown off during his weekly Saturday luncheons. These were formal affairs that featured the brightest lights from across the intellectual spectrum; they would gather for fine food and conversation that would follow according to questions Oppenheim would frame for each course. Einstein was a regular at these gatherings.

Additionally, Einstein and Oppenheim shared a deep love

for philosophy. Back in Germany he had connected Oppenheim with Hans Reichenbach, Einstein's former student and eventual colleague at the University of Berlin. Reichenbach was one of the first philosophers to understand and develop the philosophical ramifications of the theory of relativity—he had been one of the few students to attend Einstein's first class on the general theory in 1919—and Einstein tried to use his influence to secure Reichenbach a position in the philosophy department. When he failed because the philosophers at Berlin did not recognize this new philosophy of science as real philosophy, Einstein convinced Planck to create a chair for Reichenbach in the physics department, specializing in the foundations of physics.

Reichenbach supplied Oppenheim with a long string of his recent doctoral students who, because their technical combination of modern science and philosophy challenged the traditional elements of the discipline, had a difficult time finding positions in German universities. Oppenheim would give these young philosophers a small stipend, and they would work with him on developing ideas. With these ideas turned into publications they would launch their careers, and Oppenheim would move on to his next young collaborator. Oppenheim was a strong pull to Princeton for Einstein, and the two would stroll together every Sunday.

Einstein and Elsa purchased a two-level home at 112 Mercer Street, which was within walking distance of the Institute. Unfortunately, it was not a short walk, and one day not long after Einstein had joined the Institute his secretary received a phone call requesting Professor Einstein's home address. The secretary told the caller that she was not at liberty to give such personal information out over the phone, at which point the caller confided, "I *am* Dr. Einstein. Please don't tell anybody, but I'm on my way home, and I've forgotten where my house is."[6]

At Princeton he would be able to focus on his scientific

interest, which was the next step in the development of a complete single unified theory that could account for the behavior of universe as a whole. In 1905 he had published five papers. The last two were his theory of special relativity, and in 1909 he had realized the need to extend the treatment, to generalize it to account for gravitation. He had been successful in this in 1916 but realized that this generalization was not complete. His first two papers concerned the existence of atoms, and the third the nature of light. These formed a coherent whole in Einstein's mind, as the atomic hypothesis showed that matter was made up of small particles and the quantum light paper had shown the same to be true of light. If, as his theory of relativity had shown, light and matter are just different forms of energy, and this had to be understood in terms of discrete bits, then there ought to be a single theory, a single image of the universe that could be pictured in Einstein's head, that brought together the relativistic work with the quantum work. This grand unified theory had become Einstein's single-minded goal, and he would direct untold hours throughout the rest of his life in search of it. For Einstein, beauty was a mark of truth, and unity and coherence was a mark of beauty. Simplicity and elegance were achieved by intellectual consolidation, and so he worked to unify all of physics into a single comprehensive set of concepts and equations.

Einstein's interest was not widely shared at the time despite his acclaim. At first, the scientific consensus refused to accept the quantum hypothesis for light as anything more than the heuristic he had initially claimed it to be in 1905. Sure, the standard line went, we can think of light *as if* it is a particle at emission and absorption, but it really isn't. When the American physicist Arthur Compton won the Nobel Prize in 1927 for showing that when we look at light bouncing off of a small particle, the only way we can understand the resulting state of affairs is if light is treated as a particle, the world of physics grudgingly came around to Einstein's point of view. Light was to be thought of as *being* made up of, not as *if it was* made up of, particles—photons, as they came to be called.

But then physicists tried to make sense of the behavior of these photons and of the smaller and smaller bits of matter that we are able to observe. Quantum mechanics, or the laws of motion obeyed by quantum particles, seemed to be strange. The more physicists tried to make sense of it, the weirder it got.

Niels Bohr, whose work on the quantum had won him the Nobel Prize in 1922, led the charge. Bohr came from a prominent, wealthy family in Copenhagen, where he and his beloved brother Harald had been star soccer players for a premier Danish team—Niels was a goalie of some acclaim, but Harald went on to be a member of the silver medal–winning 1908 Danish Olympic team before becoming a groundbreaking mathematician. After a series of appointments at English and German universities, the Danish government decided to take advantage of having a favorite son so prominently placed in the world of theoretical physics. Bohr was given a former royal palace in Copenhagen to turn into an institute. He gathered the great young minds working on quantum mechanics, including Wolfgang Pauli, George Gamow (who would become the father of the big bang theory), and most notably Werner Heisenberg. Bohr and his young guns pressed quantum questions in every direction, working independently and together to develop a theory of the very small. The results did not thrill Einstein.

The theory that emerged had its first breakthrough in 1925 with Heisenberg, who was able to account for quantum phenomena with a cumbersome and complicated mathematical account using infinite-dimensional matrices. The success in matching observation was startling, but it was a nasty set of tools to use. This did not bother Heisenberg, who felt that if it was difficult to discover, it ought to be difficult to employ. But months later, Erwin Schrödinger, who occupied Einstein's old position in Zurich, found a new way of working it out that turned out to be completely equivalent and used nothing more than the usual sort of second-order differential equations with which physicists were well accustomed to working. Schrödinger's equation made quantum mechanical descriptions of the world much easier to explore.

But this ease belied the results that were coming out. All physical theories have particular quantities, what physicists call "state variables," that are the necessary quantities to specify if one wants to know how a physical system evolves over time. These state variables are brought together in state equations that show how they relate to each other. In Newtonian mechanics, for example, the state variables are position, mass, velocity, and acceleration. For example, in the case of a pool table, if the location of each ball is specified, along with its weight and how fast and in what direction it is moving and accelerating at any time, the values for all of those things can be determined using Newton's laws for any time, either past or future. Schrödinger's equation was to be the state equation for quantum mechanics. The problem is that we have no idea what its state variable is. The Greek letter psi, Ψ, is used to represent it, and it has the mathematical form of a wave, so physicists took to calling it the wave-function. For a while Einstein and some other physicists tried to figure out what kind of wave it is. What is the medium? What causes the wave? No realistic interpretation held up, and so Einstein called it the psi-function, refusing to attribute wavelike properties to something that wasn't actually a wave.

While the wave-function is not itself measurable, it can be expressed in terms of an observable quantity. Indeed, we are free to determine which from a wide range of measurable quantities we want to use to express psi. This comes with two results, both of which Einstein thought deeply problematic. First, the measurable quantities from which we can choose to express psi come in pairs such that the more exactly you know the value of one member of the pair, the less the other member has a value. This is the famous uncertainty principle that Heisenberg discovered by noting a strange mathematical property of his matrices. The usual pair of properties considered, but not the only one, is position and momentum. The more exactly one measures the location of a particle, the greater the uncertainty of its momentum. The word "uncertainty" is, in a sense, unfortunate because it implies a lack of knowledge, and this is a flawed interpretation. If we know a particle's location with a very high degree of accuracy, then it is not the case that we simply *know* very little about that particle's momentum; the particle does not *have* a well-defined momentum. What Heisenberg tells us is not simply about the limitations of our knowledge about the physical world, but about what properties objects in the physical world do and do not possess. When you measure the position of an object, it ceases to have an exact value for its momentum.

This is related to the second strange result of quantum mechanics. When you use Schrödinger's equation to determine psi in a particular physical situation, the result is not a single value, but a mathematical combination of every possible value—$a_1\Psi_1 + a_2\Psi_2 + \ldots + a_n\Psi_n + \ldots$, where each Ψ represents a possible value for the observable we have chosen to express psi in terms of. If, for example, we choose position, then each Ψ will be a particular location of the object. But since there are an infinite number of exact locations, the list would be infinite. The *a*-values attached to each Ψ are numbers between 0 and 1 and have the strange property that if you square them and add them all together, they always add to 1.

The first problem is that we never see any system in a combination of possible states. Every time we look at something it is in one and only one of the possible Ψ-values. So, what sense should we make of the Schrödinger equation?

Bohr, Heisenberg, and the folks at Copenhagen, along

with Born and his people in Göttingen, contended that the Schrödinger equation is the actual state equation, and the combined (physicists say "superposed") state expressed in psi is the real value when unobserved. When we do not look at something, it is spread out through all of space. But when we do observe it, the wave-function collapses to leave the object in a single location. The squares of the *a*-values are the probabilities that when we observe the particle it will be found at the location of the corresponding location of its Ψ-value. Before the system is observed it is in the combination that comes out of the Schrödinger equation, but the instant we look at the system, it collapses from this superposed state into one and only one observable value. We cannot know which one it will be, but the computed value of psi gives us the likelihood for each possibility.

What is it about our observation that alters the system? How can Schrödinger's equation be a law of nature if it holds in reality only when we are not looking, but fails the instant we check? What causes the collapse? This is known as "the measurement problem."

Einstein thought that both the Heisenberg uncertainty principle and the measurement problem showed there to be serious flaws in quantum mechanics as it was being developed. For Einstein, physics had always started with a picture, an image in his head of how things really looked, and he then developed a mathematical description for it. He pictured light quanta, he pictured atoms, he pictured relativistic alterations in space-time. Contrary to the accusations of his Nazi enemies, Einstein's physics was not typical "Jewish" bamboozlement by manipulating formal symbols removed from reality itself. Physics, for Einstein, began when you got in touch with what reality looked like all the way down. It was a principle of his deepest belief—Einstein himself called it "religious"—that the universe was made up of real things that had at every time real values for all of their properties, which in turn were completely determined by the absolute laws that governed the universe. This was the heart of his "cosmic religion."

Quantum mechanics violated this worldview in a number of ways. The Heisenberg uncertainty principle asserted that real objects will fail to have values for certain properties they possess. This couldn't be right. The interpretation of the wave-function similarly says that systems will fail to have well-defined values for observable properties, but further that the values we will observe when actually measured in the lab are not capable of being determined before we observe them. The best we can have is a probability. This really could not be right. The laws of nature do not include probability, they are completely deterministic. They give you the real values for the properties that the objects really have.

Einstein was not bothered by probabilities in general. His argument for the existence of atoms and his use of light quanta were based on probability. Similarly, he had been working with the Indian physicist Satyendra Bose on results that Bose realized came out of Einstein's work and came to be called Bose-Einstein statistics. But in all of these cases, the use of statistics was the result of having to account for large quantities of things where the interactions are so complex that we simply cannot compute the individual values for all of the particles. It is not that each particle does not have a value, it is simply that it is too hard for us to figure them out, so the best we can do is account for the large-scale spread.

But quantum mechanics was different. It wasn't that it was too complex to know, it was that there was nothing to be known. This, Einstein thought, had to be wrong. The problem for him was that for every known result and every predicted result, the new quantum theory was dead-on right. Einstein had no problem here, saying that quantum mechanics was correct as far as it went, but, he argued, it did not go far enough. Quantum mechanics is an incomplete theory. The part we have is right, but the interpretation is wrong because there is more to be known. There are other quantities, so-called hidden variables, that we need to discover, and once we do we will be able to complete the theory and restore values to all observable properties of all objects at all times and do away with the probabilities in the theory.

In lacking this, quantum mechanics was threatening the very basis of the entire scientific project—causality. The possibility of doing science in the first place requires a basic assertion that causes will always bring about the same effects. Scientists look for laws of nature which are nothing but mathematical expressions setting out what causes what. If there is no causality, there is no science. David Hume, the Scottish Enlightenment thinker, was one of Einstein's favorites, and he was reading his work just before he developed the theory of relativity. Hume had argued that we can never prove the existence of a cause-and-effect relationship, that the best we can get is "after," but never "because of." Hume asserted that the best we can do is assume it, and Einstein was fine with this, but quantum mechanics was going one step further and denying it. To Born, Einstein said, if it meant surrendering the notion of causality, "I would rather be a cobbler or a casino clerk than a physicist."[7]

To the younger theoretical physicists, Einstein's line showed that he had changed sides. He had gone from the young rebel revolutionizing science by overthrowing entrenched ideas to being an old fuddy-duddy. Einstein did not see it this way, contending that what led him to relativity and the beginning of the quantum revolution was exactly the postulates he was standing up for. He would not step aside and let progress roll, rather he would continue working on his own. He relished being the outsider, even if it made him seem reactionary.

One sure way to show Bohr and his whippersnappers the error of their ways was to give them a puzzle that showed that quantum mechanics led to contradictory results. It became a regular exercise that Einstein would propose a simple example, such as a box full of light on a scale from which one photon was released, and ask how a quantum explanation could avoid some serious problem. Over and over again, Einstein would come up with his scenarios and Bohr would stare at him and ask for some time, only to give a complete explanation the next day when Einstein would have another puzzler ready and waiting.

The most well known of these was produced in 1935, a decade into the protracted debate, and is called the Einstein-Podolski-Rosen paradox. It is straightforward to create pairs of what we call "correlated" particles—that is, a pair of, say, electrons such that if one electron has a particular value for a given property, the other will have an equal but opposite value. For example, electrons have a property called spin and we can create electron pairs so that if one has spin +1, the other must have spin –1. According to Bohr's understanding of quantum mechanics, if we create the pair but do not observe them, then both electrons must be in the superposed states of having both +1 and –1 spin. The instant we do observe one, however, it will collapse instantaneously into having a single spin value—let's say +1. But since the particles are correlated, because they always have to have equal but opposite values, as soon as one collapses and becomes +1, the other will then also have to collapse instantaneously and become –1.

Here's where Einstein got tricky. Suppose we take the two unobserved electrons and send them far away from each other: suppose we create the pair in Kansas and send them east and west and do not observe one until it reaches New York, at which time the other would be in Los Angeles. Just before the observation, the New York and Los Angeles electrons would both be in their superposed states of +1 and –1. The instant we test for spin value in New York, it becomes +1. At that same instant, the electron in Los Angeles becomes –1. Here's the question: how does the electron in LA know that the New York electron was observed? How does it know to go from its superposed state to a definite state and how does it know which definite state to occupy so it is the opposite of its twin? The electron in New York had to tell its partner what happened, but that means it had to send a signal. The theory of relativity tells us that this signal has a limiting speed—the speed of light. We know how far it is from New York to Los Angeles and we know the speed of light, so we can figure out how long it would take for the electrons to signal to one another: about a sixty-third of a second. So the question is, what happens during that time? If the Los Angeles electron remains superposed until it gets the message, then during that period there is a violation of the conservation law by which we created the correlated pair. If the Los Angeles electron has the –1 spin value, then one of two things happened—either the signal traveled faster than light, thereby violating the theory of relativity, or it had the –1 value all along, thus violating the Copenhagen account of quantum mechanics.

Einstein thought he had them with this one. No one wanted to give up on the theory of relativity and the notion of causality it was based upon, so they would have to give up the quantum nonsense. It later turned out that Einstein's worst scientific fears were realized. Experimentalists figured out a way to see if the electron was in its superposed state without collapsing it, and it turned out that it took on its value before a speed-of-light limited signal could reach it. This should have meant that Einstein was right, that it therefore must have had the value all along, but the experiment also showed that it was in fact in a superposed state beforehand. Einstein's "hidden variables" most likely did not exist.

Einstein was alienated from his home because of the political conflicts in Germany, he was alienated from other physicists because of theoretical conflicts surrounding quantum mechanics, and he was also alienated from Zionism as well. Einstein had all along called for Jewish settlements in Palestine that would exist alongside the Arabs who already lived there. He urged the creation of democratically grounded institutions to bring both sides together in thinking through questions of planning and logistics so that they could live peaceably side by side. He warned that nationalistic sentiment would lead to violence, death, and the moral undermining of the Zionist cause.

> "The first and most important necessity is the creation of a modus vivendi with the Arab people. Friction is perhaps inevitable, but its evil consequences must be overcome by organized cooperation, so that the inflammable material may not be piled up to the point of danger. The absence of contact in every-day life is bound to produce an atmosphere of mutual fear and distrust, which is favorable to such lamentable outbursts of passion as we have witnessed. We Jews must show above all that our own history of suffering has given us sufficient understanding and psychological insight to know how to cope with this problem of psychology and organization: the more so as no irreconcilable differences stand in the way of peace between Jews and Arabs in Palestine. Let us therefore above all be on our guard against blind chauvinism of any kind, and let us not imagine that reason and common-sense can be replaced with British bayonets."[8]

The phrase "British bayonets" here refers to the use of British force to put down Arab uprisings which had been occurring in Palestine as the number of Jewish immigrants increased as the economy in Europe worsened following the collapse of 1929. Einstein was urging dialogue and the arrangement of shared governance, but in letters of increasing urgency Weizmann was trying to shut down that line of argument by insisting to Einstein that the Zionists had no one with whom they could negotiate and that the violence was being perpetrated only by the Arabs on Jews. Weizmann was using these uprisings to argue for the stronger nationalistic line that Einstein so detested, and for supporting harsher British efforts to quell Arab violence against the Zionist project in Palestine.

From Einstein's perspective, this shift rightward drove too many Zionists toward the far right-wing Revisionist Party, which not only sought complete control over biblical lands, but espoused beliefs that some of the most prominent members of the party modeled on European fascism, cleansed of its antisemitism. To Einstein, who saw the rise of fascism as one of the greatest threats to humanity, the idea of Jewish fascism not only was dangerous to the occupants of Palestine, it left Judaism itself in peril. "Under the guise of nationalistic propaganda Revisionism seeks to support the destructive speculation in land; it seeks to exploit the people and deprive them of their rights. Revisionism is the modern embodiment of those harmful forces which Moses with foresight sought to banish when he formulated the model code of social law."[9]

When the Revisionist leaders objected, Einstein left no doubt that he thought their brand of nationalistic Zionism needed to be defeated. In a letter to Beinish Epstein of the Revisionist Zionist Party, Einstein wrote,

> You certainly know better than I that my characterization of the aims of the revisionist party were appropriate.
>
> They borrow methods from the Fascists that I abhor deeply, and use them to serve the interests of those who, relying on their ownership of the means of production, disenfranchise and exploit the non-owners.
>
> I am convinced that it is the duty of everyone who is serious about Zionism to fight your party with all available means.
>
> It is my opinion that [the success] of your party would signify the bankruptcy of the Zionist objectives.
>
> In my eyes, the revisionist approach to the Arabic problem is as ignoble as it is dangerous.[10]

Einstein was fighting against nationalism based on what he saw as "blind chauvinism," but he and those like him, such as

Martin Buber, were being more and more marginalized in the larger conversation.

Things were no less stormy with the Hebrew University. Einstein pictured the university becoming a center of research equal to the great universities of Europe, a home for the world's top Jewish scholars in every discipline to make great discoveries and to create great works. Such an environment would have to be nurtured and overseen by the greatest Jewish minds from around the world because they were trying to rapidly create "the intellectual atmosphere, the collaboration of scientific minds, that intellectual aura upon which the intensity of all scholarship depends and which can normally only develop gradually."[11] The task of creating the Hebrew University, not just as another institution of higher learning, but as an intellectual torch for Jews worldwide, was an extraordinary undertaking and required the best efforts and the active involvement of the brightest lights of Jewish academia.

In Einstein's mind, this meant that control over the university's operations had to be entrusted to a small group of European Jews that included, among a few others, Einstein himself, Weizmann, and Sigmund Freud. This Jewish professoriate dream team had to make the decisions if the project was to succeed. The last place control could be placed was in Jerusalem at the university itself. Palestine was a small, sparsely populated region without an intellectual infrastructure. The locals would not be qualified to create an environment that nourished top-rate scholars.

This view, of course, was not shared in Jerusalem, especially by Judah Magnes, the first president of the university, who insisted that he be given control over university affairs. What resulted was a nasty and protracted civil war over hiring, firing, official policy, and every other aspect of running the institution. Einstein objected to everything that Magnes did, and Magnes complained bitterly that he could not do his job if micromanaged by those who had no sense of the day-to-day affairs. He took political steps to insulate his power from the school's governing board, but with each move Magnes made, Einstein grew louder and more provocative, calling for his removal or at least the appointment of a dean of faculty to oversee academic matters. Einstein and Weizmann both proposed Selig Brodetsky, a Russian-born Jewish mathematician who was on the faculty at Leeds University in England.[12]

Over time, Weizmann removed himself from the affairs of the university to focus on the larger issues, and with his departure Einstein became less and less influential, having to make more and more noise to even be heard. Eventually, Einstein resigned himself to his lack of standing, and his extrascientific efforts flowed away from Zionist causes—including Hebrew University—and toward his pacifistic concerns.

But, as with the rest of his life, this too became messy. Hitler was in power in Germany, and Einstein had little doubt that this meant another war was in the offing. The Nazis were not just spewing hateful, nationalistic, violent rhetoric, they meant business. This left Einstein's pacifism in a precarious place.

Einstein had been giving speeches in which he argued in favor of conscientious objectors. If only 2 percent of all conscripted young men would refuse to serve, Einstein contended, the possibility of war would be all but eliminated. It is not clear on what calculation Einstein was basing this 2 percent figure, but he repeated it many times.

As he was preparing to leave Belgium in 1933 for America, he was summoned to an audience with the Belgian king. Two young men were being imprisoned for refusing to serve their time in the military, and the king was concerned that Einstein would take up their cause, resulting in a mass movement just as the Nazi war machine appeared to be gearing up. The king pleaded with Einstein not to make public statements in support of the objectors and even to denounce them.[13]

This put Einstein in a precarious position. On the one hand, he was a true believer in pacifism. On the other hand, he also shared the king's belief that Belgium was in imminent danger. If the Germans had been brutal to the Belgians in the First World War, there is no telling how bad it could be the second time round. To the chagrin of pacifists worldwide, Einstein complied with the king's wishes and urged the Belgian young men to serve their country in the military. Einstein argued that he was not abandoning his pacifist principles, but rather that the facts on the ground had become pathological. It is immoral, Einstein argued, to be a part of a military war machine that used violence as a tool to advance national interest. But this was not the case in Belgium. The Belgian force was serving a purely defensive purpose, trying to do nothing but protect the lives of innocents from the irrational evil of Hitler. One country should never declare war on another, and if the government does make such a declaration, it is incumbent on the good people to object and refuse to fight. But while he in general would never himself have become part of a fighting force, "under today's conditions, if I were a Belgian, I would not refuse military service, but gladly take it upon me in the knowledge of serving European civilization."[14] This is not the last time he would be forced to argue uncomfortably for military engagement against Nazi Germany in support of peace.

While his pacifist allies were unhappy with his less than perfectly pure opposition to military service, his new host in Princeton was less happy with his pacifistic work in general. Flexner loved having Einstein the scientist as the marquee name for the new Institute for Advanced Study, but Einstein's left-leaning political actions did not sit well with the wealthy donors upon whom Flexner's livelihood depended. Flexner tried direct and indirect means of getting Einstein to stop giving public talks and publishing articles about political matters. The attempts to silence him made Einstein furious and he lashed out, flexing his muscles to show that his freedom of expression was nonnegotiable. Flexner unhappily backed off.

His new life in Princeton left Einstein isolated. Hitler had driven him away from Germany and his European friends and colleagues. Quantum mechanics drove him out of mainstream physics. Zionists' increasingly nationalist stand drove him away from his efforts on behalf of Jewish Palestine. Magnes and the Palestinian faculty caused him to lose interest in Hebrew University. And now there was the attempt to marginalize him politically in America. If this was not bad enough, Einstein soon suffered personal tragedy. Not long after moving into 112 Mercer Street, Elsa became ill. Her condition rapidly deteriorated, and she passed away in 1936. His beloved sister Maja came to live with him, but she too passed away two and a half years later. Einstein had lost the people with whom he was closest and was being distanced from the ideas about which he cared the most. And then there was the war.

To Einstein the start of World War II had long seemed inevitable, and for some time he had been warning others that the Nazis were preparing to launch hostilities. As a child in Munich, he had seen the combination of nationalism, militarism, and antisemitism. Nazism was the worst of worst nightmares come true. It angered and saddened him, but it did not surprise him.

What did surprise him was a visit to a summer home he was renting on Long Island in 1939 by a couple of his colleagues, Eugene Wigner and Leo Szilárd. Szilárd told Einstein of an idea he had recently considered, one that would allow for the possibility of atomic weapons. When Einstein's second paper on the theory of relativity noted the relation between energy and mass, it was quickly realized that a tiny bit of mass could create a huge amount of energy. This energy could be used for positive, constructive purposes or it could be used for militarily destructive purposes. But the discussion was purely academic since there was no known way to actually convert mass into another form of energy and, even if it were possible, development of such a scheme would be generations away, so everyone thought.

With the advances in quantum mechanics and the interest in radioactivity, physicists and chemists began looking in much more detail at the structure of the atom. The nucleus had become a focus of interest, and, with the discovery of the neutron, things got interesting. Atoms themselves are small—one ten-millionth of an inch in diameter—and the nucleus is tiny, only about a millionth of the atom's tiny volume. So, squished into this incredibly small area is virtually all of the atom's mass, its protons and neutrons. The problem is that while neutrons are electrically neutral, protons are positive, and not only do like charges repel, but the closer they are the harder they push against each other. How, then, do the nuclei of atoms stick together? They should be blowing themselves up because of the electrical repulsion of the protons. The only explanation would be another force that is, at least at close range, stronger than the positive-on-positive repulsion. Every nucleus would have to have a binding energy that kept it stable.

As scientists studied different elements, they realized that each had a different binding energy needed to maintain nuclei of that type. It was not the simplest sort of relationship, but an interesting one in which very light and very heavy elements had higher binding energies, and those in the middle had lower ones. Joining two small hydrogen atoms into a slightly larger helium atom released some of that binding energy in other forms—that is how stars, including our own sun, give off light and heat. But very heavy elements are more likely to be radioactive, breaking down into lighter elements and emitting bits of their old, heavier selves at great speeds.

Otto Hahn and Lise Meitner were studying heavy elements. Hahn was a well-established younger physicist working in Berlin when Meitner came to the university to study under Planck from Vienna. There were few women in academia in general, but even fewer in the sciences. Meitner was brilliant and Planck—hardly a cultural progressive in any sense—recognized her intelligence and talent and agreed to take her on as an assistant. Meitner and Hahn quickly hit it off, soon referring to each other as brother and sister. Working as a team and making use of Hahn's preferential gender, they made important discoveries about the nature of radioactivity and the nucleus. Then came World War II. Not only was Meitner a woman, she was a converted Jew and so had to flee Germany.

Arriving in Sweden, she received a letter from Hahn in 1939 giving the latest details on the research they had started. Hahn had no idea what to make of the extra energy he was measuring, but Meitner realized what it was—Hahn had succeeded in splitting the atom. Einstein's $E = mc^2$ had been put into use in the laboratory under test conditions. She immediately went to Bohr with the news, and the discussion about the use of this nuclear energy began among scientists.

The conversation was not seen as urgent. Splitting the atom was hard to do, and the energy you needed to put into the system to accomplish it was more than what you reaped as a result. This limitation would need to be overcome if this discovery were to be turned into a usable technology, either peaceful or as a weapon of mass destruction. No one had any clue how to do this. Some—including Einstein—thought it could not be done. You would have to knock something against a nucleus hard enough to get it to break into pieces. Nuclei were small and electrically charged. No matter what sort of "bullet" you shot at them, you would miss almost all of the time. And the bullets would all have to be shot at an energy high enough to break apart the nucleus just in case it happened to be the shot that hit the invisible target. That is a lot of wasted energy spent in the hopes of getting lucky once.

A few years earlier, Szilárd pondered this problem while stopped at a red light in heavy London traffic. The fact that nuclei were made up of protons meant that they had a positive charge. Electrons are too small to do the job as bullets, and protons would be repelled by the nucleus's positive charge. The only thing that would work would be a neutral particle of significant size—neutrons. But then we have the problem that Einstein described as "like shooting birds in the dark in a place where there are not many birds."[15] You would need lots and lots of shots. An idea occurred to Szilárd. With some forms of larger elements, when the nucleus breaks apart into smaller pieces, the resulting atoms have fewer neutrons in their nuclei than are in the original larger atom. The liberated neutrons have to go somewhere, and they would take some of the freed up binding energy in the form of the speed of their exit. In some atoms, the exploded nucleus gives up two or three fast-moving neutrons, which would then become bullets. One successful bullet into a nucleus would cause the shot target to then become two or three guns, shooting even more bullets at the surrounding atoms, which could then shoot even more. Hit one nucleus and it could hit three, which could then hit nine, which could then hit twenty-seven, which could then . . . Szilárd had conceived of the possibility of a nuclear chain reaction.

When Szilárd heard about Hahn's results and the work by researchers at Cambridge using streams of highly energized particles, he realized that uranium would be the fuel of choice, and he then recalled that Belgian colonies in Africa held vast reserves of uranium, reserves that needed to be kept away from the Nazi government and their researchers. He needed to tell the Belgians at the highest level that they needed to step in. But how could he get to the Belgian power elite? Before Princeton, Einstein had been in Belgium and had become close with the Belgian royal family. Einstein could get a note to the Queen of Belgium.

In 1939 Szilárd went to Wigner, who was in Princeton, and they went to see Einstein. But he was out of town for the summer. They got word from Einstein's secretary that he was on Long Island, renting a house from a Dr. Moore. She didn't have an address but did know the town—Peconic. The two jumped in Wigner's car and sped off. Arriving in Peconic, they asked everyone they encountered for the address of Dr. Moore's house, but to no avail. Then they realized they were asking the wrong question. Stopping a boy on the side of the road, they asked where Dr. Einstein was staying and immediately got directions.

Einstein was delighted to see his colleagues. But when Szilárd told him about his idea concerning chain reactions in uranium and the possibility of German control of African uranium, he realized the gravity of the situation and thought that Szilárd's suggestion of writing the Belgian royal family was a good one. Wigner was concerned that Einstein's notorious political views and the status of all three of them as refugees might make direct contact problematic and wondered if they should first contact the U.S. State Department to get approval, which would not only give legitimacy from the American perspective, but also lend the note more weight in convincing the Belgians to act.

Now, how to get to the State Department? Initially, they considered the most politically connected figure that any of them knew personally. Einstein had met Charles Lindbergh at a couple of events. Lindbergh was the son of a congressman and a highly influential figure. Unfortunately, he was also a Nazi sympathizer urging America isolationism. They decided against approaching the aviator and instead contacted the well-connected economist Alexander Sachs through mutual friends. Sachs was a friend of President Roosevelt's and suggested that the communiqué should go directly to him. Einstein and Szilárd agreed. Wigner had departed for a prearranged trip to California, so the remaining two scientists were joined by another refugee physicist, Edward Teller. Einstein dictated the letter in German to Teller, who then gave it to Szilárd to translate. Szilárd worked up two versions and sent them both to Einstein, who signed one and sent it to Sachs.

Sachs did not immediately go to Roosevelt, and the delay concerned Einstein and Szilárd, but when he was able to secure a proper meeting, he read the note aloud to FDR. This was not the first time Nazi development of atomic weapons had been considered by the American government, but it was decided at that time that action was required, which resulted in panels and committees investigating the possibility of weaponizing nuclear fission. Ultimately, this work would give rise to the Manhattan Project, which would, without help from Einstein, create the first nuclear bombs.

Einstein was not invited to be a part of the Project because of concerns about his politics. The FBI, which had long kept a file on Einstein, considered him subversive for his political, especially his pacifistic, activities.[16] Nevertheless, he had no interest in being part of the group. This is not to say that Einstein made no contribution to the American war effort; he was used as a consultant by the navy to investigate some interesting properties of underwater explosives,[17] but none of this brought him anywhere near the Manhattan Project.

The Project, of course, was successful where the German program was not. The Germans were defeated and Einstein was as happy as anyone. Yet the work on the atomic bomb continued unabated despite the lack of a Nazi threat. Szilárd went to Einstein for another note, this time to President Harry Truman. Szilárd was now a part of the Manhattan Project and so could not tell Einstein everything he knew, but they wrote another note to the White House, urging the work be halted or at least scaled back.[18] Of course, it was not and the bombs were built, tested, and dropped on Japanese cities.

Einstein was horrified by the killings. A popular but unverified quotation has been attributed to him on hearing of the dropping of the bomb on Hiroshima. He supposedly said, "I could burn my fingers that I wrote that letter to Roosevelt." Whether he spoke those words or not, it is clear that Einstein was concerned about the reality of nuclear weapons. In a talk in 1945 he declared that "the war is won, but the peace is not":

> "Physicists find themselves in a position not unlike that of Alfred Nobel. Alfred Nobel invented the most powerful explosive ever known up to his time, a means of destruction par excellence. In order to atone for this, in order to relieve his human conscience he instituted his awards for the promotion of peace and for achievements of peace. Today, the physicists who participated in forging the most formidable and dangerous weapon of all times are harassed by an equal feeling of responsibility, not to say guilt. And we cannot desist from warning, and warning again, we cannot and should not slacken in our efforts to make the nations of the world, and especially their governments, aware of the unspeakable disaster they are certain to provoke unless they change their attitude toward each other and toward the task of reshaping the future. We helped in creating this new weapon in order to prevent the enemies of mankind from achieving it ahead of us, which, given the mentality of the Nazis, would have meant inconceivable destruction and the enslavement of the rest of the world. We delivered this weapon into the hands of the American and British people as trustees for the whole of mankind, as fighters for peace and liberty. But so far, we fail to see any guarantee of peace, we do not see any guarantee of the freedoms that were promised to the nations in the Atlantic Charter. The war is won, but the peace is not."[19]

The nuclear age had begun, and if the world returned to the old nationalistic model, the results could be horrific.

Although Einstein spoke for many scientists, he did not reflect the sentiments of all of them. Indeed, Edward Teller, who took down Einstein's words for Roosevelt, felt no guilt and thought that the whole endeavor needed to continue moving forward, becoming the father of the hydrogen bomb. Teller's work worried Einstein, and he appeared as the inaugural guest on the television program *Today with Mrs. Roosevelt* on February 12, 1950, where he warned that an atomic arms race with the Soviet Union could lead to unthinkable consequences. Having won World War II with the bomb, the illusion is now commonplace among Americans that "security could be won by permanent and decisive military superiority."[20] But this is not true. It would lead only to financial and psychological ruin.

> The arms race between the United States and the Soviet Union, initiated originally as a preventive measure, assumes hysterical proportions. On both sides, means of mass destruction are being perfected with feverish haste and behind walls of secrecy. And now the public has been advised that the production of the hydrogen bomb is the new goal which will probably be accomplished. An accelerated development toward this end has been solemnly proclaimed by the President. If these efforts should prove successful, radioactive poisoning of the atmosphere and, hence, annihilation of all life on earth will have been brought within the range of what is technically possible. The weird aspect of this development lies in its apparently inexorable character. Each step appears as the inevitable consequence of the one that went before. And at the end, looming ever clearer, lies general annihilation."[21]

The show was widely seen and well received, but there was one viewer who deeply disliked what he saw—FBI director J. Edgar Hoover, who had been keeping a file on Einstein. The scientist had his share of right-wing detractors in America as well as Germany. When Einstein first started spending time at Cal Tech, a group called the "Woman Patriot Corporation" worked up an extended document accusing Einstein of being a

Communist, an anarchist, and a menace who should not be allowed on U.S. soil. They accused him of being associated with more "anarcho-communist groups" than Josef Stalin.[22] The document ended up in Hoover's files.

But it was the appearance with Mrs. Roosevelt that moved Einstein to the top of Hoover's list. Hoover, whose hatred was so broad and deep as to know few boundaries, loathed Eleanor Roosevelt like few others. Einstein's appearance and his message marked him, and Hoover not only collected information, but began an active campaign to smear Einstein and undermine his international reputation, the ultimate goal being his deportation.[23]

Hoover had his people working hard to come up with a smoking gun he could use to connect Einstein to the Communist menace. It seemed promising when a man walked into the Miami office of the FBI in September of 1953 claiming to have information about Einstein's long-time Communist associations going back to Berlin, where he personally knew Einstein. A memo following the interview was immediately sent to the Newark office, which was coordinating the anti-Einstein efforts. "An individual of German origin who has been in the United States but a few years, stated that he has known of [Einstein] for many years and that in 1905 EINSTEIN developed a hypothetical theory of relativity . . . in 1919 . . . the German Left-wing press hailed EINSTEIN as a great scientist . . . [However] Einstein . . . was not a scientist or a philosopher, but was a politician who would bring the German people to anarchism and Communism."[24] This informant claimed to have the goods on Einstein, and that Hoover could have him arrested or deported. Although the name was classified for decades, scholars figured it out, and later the declassified file confirmed that the mystery informant was none other than Paul Weyland, the partner of Ernst Gehrke, who created the original anti-Einstein circus in Berlin.

Although Hoover did all he could to bring down Einstein, the scientist's persona and reputation were different from those of so many whom the FBI director was able to destroy during the McCarthy era. Einstein may not have been influential enough to bring about the major structural changes he thought were necessary to securing world peace, but he was allowed to be Albert Einstein for the last years of his life.

Had Einstein been ordered to leave, he would have had a secure place to go. In November 1952, after the death of Chaim Weizmann, who had been made the first president of Israel, David Ben-Gurion, the Israeli prime minister, decided that the post would be offered to Einstein. Ben-Gurion did not think Einstein the best candidate for the office given his politics, but felt that there was no other possibility than to offer it to him. The Israeli ambassador, Abba Eban, was ordered to make the offer, but whether it was speculation or an authentic leak that reached Einstein, he received advanced word of the pending offer. Einstein was in a panic. He was unqualified, but could he refuse? He phoned Washington and reached Eban before he left for Princeton, catching him completely off guard. Einstein said that he was tremendously honored, but that the offer should not be made. Needing a formal reply, Eban had a deputy take an official offer to Princeton, where he received Einstein's ready response.

> I am deeply moved by the offer from our state of Israel, and at once saddened and ashamed that I cannot accept it. All my life I have dealt with objective matters, hence I lack both the natural aptitude and experience to deal properly with people and to exercise official functions. For these reasons alone I would be unsuited to fulfill the duties of that high office, even if advancing age was not making increasing demands on my strength.
>
> I am the more distressed over these circumstances because my relationship to the Jewish people has become my

> strongest human bond, ever since I became fully aware of our precarious situation among the nations of the world.[25]

Einstein was relieved to have the matter over, but more relieved was Ben-Gurion, who asked his assistant Yitzak Navon, "Tell me what to do if he says yes! I've had to offer the post to him because it's impossible not to. But if he accepts we're in for trouble!"[26]

Einstein's last years were spent in Princeton, in the company of his secretary Helen Dukas, writing and speaking for pacifistic causes, working to help Jewish refugees flee Europe for the safety of the United States, and developing his grand unified theory.

His house was close to the African-American section of Princeton, then a very segregated town, and Einstein would regularly walk through the parts of town that rarely saw white people. Einstein was different. He would stroll along, talking with people relaxing on their porches, joking with the kids and handing them nickels or candy, even regularly helping one young girl with her arithmetic homework (for which she would give him gumdrops).[27] He was loved there as everywhere else, but to them he was different. He was remembered by people in the neighborhood as "a friend of black people." "He was comfortable [in the African-American section of town]"; "there was just no racism about him."[28] Early on from his time in the United States, Einstein recognized the treatment of African-Americans in America—it was the treatment of Jews in Europe. With his black neighbors, as with so many oppressed people across the globe, he felt a kinship.

It was with people like this that he felt the closest to home, a home he rarely had. Einstein renounced his German citizenship as a young man, never having felt Munich to be his home. He loved his time in Italy, but always as a foreigner. Switzerland, with his adopted family the Wintelers and especially in cosmopolitan Zurich, was one place he did feel he belonged. But Prague, Berlin, Pasadena, and Princeton were just places he lived, more or less pleasant, but never home. He did become an American citizen, living out his last years walking the streets of Princeton, a local fixture and global celebrity. He still wrote, he still worked on his physics, and he still thought about the universe, the world, and the neighborhood.

Einstein had never been particularly healthy. He suffered from stomach ailments most of his life, and several times he was certain he would die from them. In 1955, Einstein was seventy-six. He had an aortic aneurism. On April 13 he experienced severe pain. It was clear that this would be the end, and Einstein refused to accept any sort of medical treatment that would artificially delay his passing. It was "tasteless," he said. Hans Albert flew in from California, and his friends gathered. Two days later he was hospitalized. For a while he stabilized, speaking and joking with visitors and hospital staff. On Monday, April 17, Einstein awoke at one o'clock in the morning. His aneurism burst, and he died. Moments earlier, he uttered a single sentence aloud in his native tongue. Einstein's dying words were heard by one person, a night nurse who spoke no German. In death as in life, Albert Einstein left us a mystery.

NOTES

Introduction

1. Quoted in Clark, *Life and Times*, 258.

2. Letter to Max Born, September 9, 1920, reproduced in *Einstein/Born Correspondence*, 35.

3. Quoted widely; see for example Dooling, *Rapture*, 137, but without a clear source.

Chapter 1. Everything Was in Order

1. Stachel, *CPAE*, Vol. 1, xvi.

2. Reiser, *A Biographical Portrait*, 28.

3. Quoted in Frank, *Life and Times*, 14.

4. Stachel, *CPAE*, Vol. 1, xix.

5. Einstein, Autobiographical Notes, in Schilpp, *Albert Einstein, Philosopher-Scientist*, 9.

6. Davis et al., *The Mathematical Experience*, 188.

7. Grandin, *Thinking in Pictures*, 180–82.

8. Muir, "Einstein and Newton Showed Signs of Autism," 10.

9. Einstein, Autobiographical Notes, in Schilpp, *Albert Einstein, Philosopher-Scientist*, 15.

10. Quoted in Clark, *Life and Times*, 13.

11. Frank, *His Life and Times*, 11.

12. Quoted in Jammer, *Einstein and Religion*, 23–24.

13. Einstein, Autobiographical Notes, in Schilpp, *Albert Einstein, Philosopher-Scientist*, 9.

14. Ibid.

15. Ibid., 3–5.

16. Clark, *Life and Times*, 20.

17. Isaacson, *His Life and Universe*, 23.

18. Stachel, *CPAE*, Vol. 1, xxii.

19. Letter to Mileva Maric, 16,2, 1898, in Renn and Schulmann, *The Love Letters*, 5.

20. Letter to Mileva Maric, 13, 3, 1899, in Renn and Schulman, *The Love Letters*, 8.

21. Letter to Mileva Maric, 10, 8, 1899, in Renn and Schulman, *The Love Letters*, 11.

22. Letter to Mileva Maric, 29, 7, 1900, in Renn and Schulman, *The Love Letters*, 19.

23. Letter to Mileva Maric, 30, 4, 1901, in Renn and Schulman, *The Love Letters*, 46.

Chapter 2. The Miracle Year

1. Kuhn, *Structure*, 58.

2. Rigden, *Standard*, 43–44.

3. See Gribbin and Gribbin, *Annus Mirabilis*, 32–33.

4. Isaacson, *His Life and Universe*, 71.

5. Letter to Mileva Maric, 19, 12, 1901, in Renn and Schulmann, *The Love Letters*, 71.

6. Seelig, *Albert Einstein*, 112.

7. Quoted in Fölsing, *A Biography*, 124.

8. Stachel, *CPAE*, Vol. 2, 123.

9. Ibid., 134.

10. Gribbin and Gribbin, *Annus Mirabilis*, 77–78.

11. Poincaré discusses his formulation of the principle of rela-

tivity in *Value of Science*. For an in-depth discussion of the relation between Poincaré's work and Einstein's, see Galison, *Einstein's Clocks, Poincaré's Maps*.

12. Einstein et al., *The Principle of Relativity*, 63–64.
13. Ibid., 70.
14. Ibid.
15. Newton, *Principia*, 6.

Chapter 3. The Happiest Thought

1. Quoted in Pais, *Subtle Is the Lord*, 150.
2. Ibid.
3. Letter to Laub, quoted in Clark, *Life and Times*, 111.
4. Quoted in Fölsing, *A Biography*, 203.
5. Einstein, Autobiographical Notes, in Schilpp, *Albert Einstein, Philosopher-Scientist*, 15.
6. See Reid, *Hilbert*, 1970, for a wonderful discussion of Minkowski and his relationship with Hilbert.
7. Einstein et al., *The Principle of Relativity*, 75.
8. Pais, *Subtle Is the Lord*, 152.
9. Quoted in Clark, *Life and Times*, 122.
10. Ibid.
11. Ibid., 205.
12. Fölsing, *A Biography*, 229.
13. Ibid., 250.
14. Levenson, *Einstein in Berlin*, 100.
15. Frank, *His Life and Times*, 76.
16. Ibid., 77.
17. Ibid.
18. Ibid., 84.
19. Letter to H. Zangger, quoted in Fölsing, *A Biography*, 287.
20. Pais, *Subtle Is the Lord*, 213.
21. Fölsing, *A Biography*, 318.
22. Reid, *Hilbert*, 141–42.
23. For a detailed account of Einstein's wrestling with the hole problem, see Norton, "Einstein, the Hole Argument, and the Reality of Space."

Chapter 4. Two Wars

1. Frank, *His Life and Times*, 120.

2. Quoted in Fölsing, *A Biography*, 350.

3. Rosenthal-Schneider and Braun, *Reality and Scientific Truth*, 74.

4. The most comprehensive account of Einstein's long march to the Nobel Prize is in Elzinga, *Einstein's Nobel Prize*. See also Pais, *Subtle Is the Lord*, 502–11.

5. Address to the French Philosophical Society, April 6, 1922, quoted in Calaprice, *Expanded Quotable Einstein*, 9.

Chapter 5. The Worldwide Jewish Celebrity

1. Fölsing, *A Biography*, 273.

2. Isaacson, *His Life and Universe*, 243.

3. Levenson, *Einstein in Berlin*, 118–223.

4. Fölsing, *A Biography*, 460.

5. Ibid., 462.

6. Clark, *Life and Times*, 257.

7. Translated in Hentschel, *Physics and National Socialism*, 1.

8. Ibid.

9. Ibid.

10. Quoted in Clark, *Life and Times*, 264.

11. Levenson, *Einstein in Berlin*, 255.

12. For an in-depth account of Haber's life, see Fritz Stern's *Einstein's Jewish World.*

13. Stachel, *CPAE*, Vol. 9, 368.

14. Einstein, "Why Do They Hate the Jews?," in *Ideas and Opinions*, 194.

15. Ibid.

16. Einstein, "Our Debt to Zionism," in *Ideas and Opinions*, 190.

17. Einstein, "Letter to Professor Dr. Hellpach, Minister of State," in *Ideas and Opinions*, 171.

18. Einstein, "How I Became a Zionist" in Jerome, *Einstein on Israel and Zionism*, 45.

19. Einstein, "Why Do They Hate the Jews?," in *Ideas and Opinions*, 195.

20. Quoted in Jerome, *Einstein on Israel and Zionism*, 25.

21. Born to Einstein, October 7, 1920, in *The Einstein-Born Correspondences*, 40.

22. Quoted in Fölsing, *A Biography*, 500.

23. For an in-depth discussion of this battle and Einstein's place in it, see Rosenkranz, *Einstein Before Israel.*

24. Quoted in Stachel, *Einstein from B to Z*, 64.

25. Clark, *Life and Times*, 278.

26. Rosenkranz, *Einstein Before Israel*, 95.

27. Letter to Pauline Einstein, 10/8/1918, in Stachel, *CPAE*, Vol. 8, 906.

28. Levenson, *Einstein in Berlin*, 264.

29. Rosenkranz, *Einstein Before Israel*, 143.

30. Quoted in ibid., 159.

31. Quoted in Fölsing, *A Biography*, 549.

32. Ibid., 531.

33. Rosenkranz, *Einstein Before Israel*, 146.

34. Ibid., 151.

35. Quoted in Fölsing, *A Biography*, 531.

36. For a detailed discussion of the notion of "Jewish Science" and Einstein's reaction to it, see Gimbel, *Einstein's Jewish Science.*

37. *German Physics in Four Volumes*, in Hentschel, *Physics and National Socialism*, 100.

38. Walker, *Nazi Science*, 6.

39. Frank, *His Life and Times*, 221.

40. Calaprice, *Ultimate Quotable Einstein*, 502.

Chapter 6. In Exile

1. Quoted in Fölsing, *A Biography*, 661.

2. Ibid., 663.

3. Einstein to Haber, August 8, 1933, in Stachel, *CPAE*, Vol. 12, 388.

4. Clark, *Life and Times*, 430.

5. For an account of the founding of the Institute, see Ed

Regis's *Who Got Einstein's Office?: Eccentricity and Genius at the Institute for Advanced Study.*

6. Isaacson, *His Life and Universe,* 439.

7. Einstein to Born, April 19, 1924, in *The Born-Einstein Letters,* 118.

8. Einstein, "Jew and Arab" in Jerome, *Einstein on Israel and Zionism,* 72–73.

9. Einstein, "The Goal of Jewish-Arab Amity," in Jerome, *Einstein on Israel and Zionism,* 104.

10. Letter to B. Epstein, April 23, 1935, quoted in Rosenkranz, *Einstein Before Israel,* 105.

11. Letter to O. Warburg, January 1, 1926, quoted in Rosenkranz, *Einstein Before Israel,* 191.

12. For an in depth discussion of Einstein's battles with Magnes, see Rosenkranz, *Einstein Before Israel,* ch. 5.

13. Fölsing, *A Biography,* 674–76.

14. Quoted in Fölsing, *A Biography,* 675.

15. Quoted in Nathan and Norden, *Einstein on Peace,* 290.

16. For an extensive discussion of the FBI's activities with respect to Einstein, see Jerome's *The Einstein File.*

17. Clark, *Life and Times,* 571–73.

18. Ibid., 583.

19. Reprinted in Rowe and Schulmann, *Einstein on Politics,* 381–82.

20. Ibid., 403.

21. Ibid., 403–4.

22. Isaacson, *His Life and Universe,* 399–400.

23. See Jerome, *The Einstein File,* for a full accounting.

24. Quoted in Jerome, *The Einstein File,* 207.

25. Quoted in Clark, *Life and Times,* 733.

26. Ibid.

27. See Jerome and Taylor, *Einstein on Race and Racism,* for extended discussion.

28. Jerome and Taylor, *Einstein on Race and Racism,* 35–37.

BIBLIOGRAPHY

Braun, Reiner, and David Krieger. *Einstein—Peace Now!: Visions and Ideas.* Weinheim: Wiley-VCH, 2005.

Calaprice, Alice. *The Expanded Quotable Einstein.* Princeton: Princeton University Press, 2000.

Clark, Ronald. *Einstein: The Life and Times.* New York: World Publishing, 1971.

Davis, Philip J., and Reuben Hersh. *The Mathematical Experience.* Boston: Birkhäuser, 1981.

Dooling, Richard. *Rapture for the Geeks: When AI Outsmarts IQ.* New York: Random House, 2009.

Einstein, Albert. *Einstein on Humanism.* New York: Carol Pub. Group, 1993; 1950.

———. *Ideas and Opinions. Based on Mein Weltbild.* New York: Dell, 1973. Laurel-Leaf Library.

Einstein, Albert, Max Born, and Hedwig Born. *The Born-Einstein Letters; Correspondence Between Albert Einstein and Max and Hedwig Born from 1916–1955.* New York: Walker, 1971.

Einstein, Albert, Otto Nathan, and Heinz Norden. *Einstein on Peace.* New York: Simon and Schuster, 1960.

Elzinga, Aant. *Einstein's Nobel Prize: A Glimpse Behind Closed Doors—The Archival Evidence.* 6 Vol. Sagamore Beach, MA: Science History Publications/USA, 2006.

Fölsing, Albrecht. *Albert Einstein: A Biography.* New York: Viking, 1997.

Frank, Philip. *Einstein: His Life and Times.* Cambridge, MA: Da Capo Press, 2002; 1953.

Galison, Peter. *Einstein's Clocks, Poincaré's Maps.* New York: Norton, 2003.

Gimbel, Steven. *Einstein's Jewish Science: Physics at the Intersection of Politics and Religion.* Baltimore: Johns Hopkins University Press, 2012.

Grandin, Temple. *Thinking in Pictures: And Other Reports from My Life with Autism.* New York: Vintage Books, 1996.

Gribbin, John R., and Mary Gribbin. *Annus Mirabilis: 1905, Albert Einstein, and the Theory of Relativity.* New York: Chamberlain Bros., 2005.

Hentschel, Klaus. *Physics and National Socialism: An Anthology of Primary Sources.* 18 Vol. Basel, Boston: Birkhäuser Verlag, 1996. Science Networks Historical Studies.

Hoffmann, Banesh, and Helen Dukas. *Albert Einstein: Creator and Rebel.* New York: Viking, 1972.

Illy, József. *Albert Meets America: How Journalists Treated Genius During Einstein's 1921 Travels.* Baltimore: Johns Hopkins University Press, 2006.

Isaacson, Walter. *Einstein: His Life and Universe.* New York: Simon and Schuster, 2007.

Jammer, Max. *Einstein and Religion: Physics and Theology.* Princeton, NJ: Princeton University Press, 1999.

Jerome, Fred. *Einstein on Israel and Zionism: His Provocative Ideas About the Middle East.* New York: St. Martin's Press, 2009.

———. *The Einstein File: J. Edgar Hoover's Secret War Against the World's Most Famous Scientist.* New York: St. Martin's Press, 2002.

Jerome, Fred, and Rodger Taylor. *Einstein on Race and Racism.* New Brunswick, NJ: Rutgers University Press, 2005.

Kuhn, Thomas. *The Structure of Scientific Revolutions.* Chicago: University of Chicago Press, 1950.

Levenson, Thomas. *Einstein in Berlin.* New York: Bantam Books, 2003.

Lorentz, H. A., et al. *The Principle of Relativity: A Collection of Original Memoirs on the Special and General Theory of Relativity.* London: Methuen and Co. Ltd, 1923.

Miller, Arthur I. *Albert Einstein's Special Theory of Relativity: Emergence (1905) and Early Interpretation, 1905–1911.* Reading, MA: Addison-Wesley Pub. Co., Advanced Book Program, 1981.

Muir, Hazel. "Einstein and Newton Showed Signs of Autism." *New Scientist* 2393. May 3 (2003): 10.

Nathan, Otto, and Heinz Norden. *Einstein on Peace.* New York: Schocken, 1960.

Neffe, Jürgen. *Einstein: A Biography.* New York: Farrar, Straus, and Giroux, 2007.

Newton, Isaac. *Mathematical Principles of Natural Philosophy.* Trans. Andrew Motte and Florian Cajori. Berkeley, CA: University of California Press, 1934.

Norton, John. "Einstein, the Hole Argument and the Reality of Space," in John Forge (ed.), *Measurement, Realism and Objectivity.* Dordrecht: Reidel, 1987. 153–188.

Pais, Abraham. *Subtle Is the Lord: The Science and the Life of Albert Einstein.* New York: Oxford University Press, 1982.

Poincaré, Henri. *The Value of Science.* New York: Dover, 1958.

Regis, Edward. *Who Got Einstein's Office?: Eccentricity and Genius at the Institute for Advanced Study.* Reading, MA: Addison-Wesley, 1987.

Reid, Constance. *Hilbert.* London, New York: Springer-Verlag, 1970.

Reiser, Anton. *Albert Einstein: A Biographical Portrait.* New York: Macmillan, 1930.

Renn, Jürgen, and Robert Schulmann. *Albert Einstein/Mileva Marić—The Love Letters.* Princeton, NJ: Princeton University Press, 1992.

Rigden, John S. *Einstein 1905: The Standard of Greatness*. Cambridge, MA: Harvard University Press, 2005.

Rosenkranz, Ze'ev. *Einstein Before Israel: Zionist Icon or Iconoclast?*. Princeton, NJ: Princeton University Press, 2011.

Rosenthal-Schneider, Ilse, and Thomas Braun. *Reality and Scientific Truth: Discussions with Einstein, Von Laue, and Planck*. Detroit: Wayne State University Press, 1980.

Rowe, David E., and Robert J. Schulmann. *Einstein on Politics: His Private Thoughts and Public Stands on Nationalism, Zionism, War, Peace, and the Bomb*. Princeton, NJ: Princeton University Press, 2007.

Schilpp, Paul Arthur. *Albert Einstein, Philosopher-Scientist*. New York: Harper and Row, 1959.

Seelig, Carl. *Albert Einstein*. London: Staples Press, 1956.

Stachel, John J. *Einstein from "B" to "Z."* Vol. 9. Boston: Birkhäuser, 2002.

———. *The Collected Papers of Albert Einstein*. Princeton, NJ: Princeton University Press, 1987.

Stern, Fritz Richard. *Einstein's German World*. Princeton, NJ: Princeton University Press, 1999.

Walker, Mark. *Nazi Science: Myth, Truth, and the German Atomic Bomb*. New York: Plenum Press, 1995.

Wittgenstein, Ludwig. *Tractatus Logico-Philosophicus*. London, New York: K. Paul, Trench, Tubner; Harcourt Brace, 1933.

INDEX

Jewish Lives is a major series of interpretive biography designed to illuminate the imprint of Jewish figures upon literature, religion, philosophy, politics, cultural and economic life, and the arts and sciences. Subjects are paired with authors to elicit lively, deeply informed books that explore the range and depth of Jewish experience from antiquity through the present.

Jewish Lives is a partnership of Yale University Press and the Leon D. Black Foundation.

Ileene Smith is editorial director. Anita Shapira and Steven J. Zipperstein are series editors.

PUBLISHED TITLES INCLUDE:

Ben-Gurion, by Anita Shapira
Bernard Berenson: A Life in the Picture Trade, by Rachel Cohen
Sarah: The Life of Sarah Bernhardt, by Robert Gottlieb
Leonard Bernstein, by Allen Shawn
David: The Divided Heart, by David Wolpe
Moshe Dayan: Israel's Controversial Hero, by Mordechai Bar-On
Einstein, by Steven Gimbel
Becoming Freud: The Making of a Psychoanalyst, by Adam Phillips
Emma Goldman: Revolution as a Way of Life, by Vivian Gornick
Hank Greenberg: The Hero Who Didn't Want to Be One, by Mark Kurlansky
Lillian Hellman: An Imperious Life, by Dorothy Gallagher
Jabotinsky: A Life, by Hillel Halkin
Jacob: Unexpected Patriarch, by Yair Zakovitch
Franz Kafka: The Poet of Shame and Guilt, by Saul Friedländer
Rav Kook: Mystic in a Time of Revolution, by Yehudah Mirsky
Primo Levi: The Matter of a Life, by Berel Lang
Moses Mendelssohn: Sage of Modernity, by Shmuel Feiner
Walter Rathenau: Weimar's Fallen Statesman, by Shulamit Volkov
Mark Rothko, by Annie Cohen-Solal
Solomon: The Lure of Wisdom, by Steven Weitzman
Leon Trotsky: A Revolutionary's Life, by Joshua Rubenstein

FORTHCOMING TITLES INCLUDE:

Rabbi Akiva, by Barry Holtz
Irving Berlin, by James Kaplan
Hayim Nahman Bialik, by Avner Holtzman
Léon Blum, by Pierre Birnbaum
Louis Brandeis, by Jeffrey Rosen
Martin Buber, by Paul Mendes-Flohr
Benjamin Disraeli, by David Cesarani
Bob Dylan, by Ron Rosenbaum
George Gershwin, by Gary Giddins
Allen Ginsberg, by Edward Hirsch
Peggy Guggenheim, by Francine Prose
Ben Hecht, by Adina Hoffman
Heinrich Heine, by Fritz Stern
Theodor Herzl, by Derek Penslar
Jesus, by Jack Miles
Groucho Marx, by Lee Siegel
Moses, by Avivah Zornberg
J. Robert Oppenheimer, by David Rieff
Marcel Proust, by Benjamin Taylor
Jerome Robbins, by Wendy Lesser
Julius Rosenwald, by Hasia Diner
Jonas Salk, by David Margolick
Steven Spielberg, by Molly Haskell
Barbara Streisand, by Neal Gabler
The Warner Brothers, by David Thomson
Ludwig Wittgenstein, by Anthony Gottlieb